T0156032

Lecture Notes in Mathematics

Volume 2322

This series reports on new developments in all areas of mathematics and their applications - quickly, informally and at a high level. Mathematical texts analysing new developments in modelling and numerical simulation are welcome. The type of material considered for publication includes:

1. Research monographs
2. Lectures on a new field or presentations of a new angle in a classical field
3. Summer schools and intensive courses on topics of current research.

Texts which are out of print but still in demand may also be considered if they fall within these categories. The timeliness of a manuscript is sometimes more important than its form, which may be preliminary or tentative.

Titles from this series are indexed by Scopus, Web of Science, Mathematical Reviews, and zbMATH.

Mohammed Hichem Mortad

The Fuglede-Putnam Theory

 Springer

Mohammed Hichem Mortad
Department of Mathematics
University of Oran 1
Oran, Algeria

ISSN 0075-8434   ISSN 1617-9692   (electronic)
Lecture Notes in Mathematics
ISBN 978-3-031-17781-1   ISBN 978-3-031-17782-8   (eBook)
https://doi.org/10.1007/978-3-031-17782-8

This Springer imprint is published by the registered company Springer Nature Switzerland AG
The registered company address is: Gewerbestrasse 11, 6330 Cham, Switzerland

Preface

When I took a course on spectral theory some 25 years ago, the Fuglede theorem was not part of the syllabus of that course. Indeed, that course was mainly directed toward unbounded self-adjoint operators, which is the main application of unbounded operators in quantum mechanics, say. The first time I heard of the Fuglede theorem was during a Banach algebra conference in honor of Barry Johnson, that took place at Newcastle University upon Tyne (U.K.) in 2001.

As readers are probably aware, the spectral theorem is the most crucial result in operator theory. We can say, at least as far as normal operators are concerned, that the Fuglede-Putnam theorem is the second salient result.

Many mathematicians have written papers about generalizations and applications of the Fuglede-Putnam theorem. Most of these contributions are included in this book. The Fuglede-Putnam theorem is not just some very useful theorem anymore, it has evolved into a small theory, and that is how I have chosen the title of this manuscript.

The present monograph is mainly intended for postgraduate students as well as researchers interested in operator theory. Readers are therefore expected to have a good knowledge of both bounded and unbounded operator theory, and while most definitions and notations should be known to them, they should read Appendix A.

Although the main topic is the Fuglede-Putnam theorem and its variants, applications, and other related results should interest readers. This book could also be taught as a specialized course. It is divided into several chapters, each chapter having its particularity. Also, most chapters have a counterexamples section in which readers may find a fair amount of interesting examples demonstrating the boundaries of possible assertions. Overall, I have tried to gather all (?) about the Fuglede-Putnam theorem as regards historical notes, classical versions with their different proofs, generalization to Banach spaces, asymptotic versions, generalizations to non-normal operators, generalizations to unbounded operators, counterexamples, applications, intertwining relations, and conjectures. My aim has not been to just list versions and generalizations of the Fuglede-Putnam theorem but to also give ideas of how to apply this powerful tool in some situations in operator theory. Sometimes, we think that a priori, there is no room for the Fuglede-Putnam theorem, yet we find an efficient place where it fits perfectly.

A comment about applications needs, however, to be singled out. There is a chapter devoted to applications that, alas, cannot all be included for obvious reasons. Indeed, I certainly cannot be aware of all applications. Moreover, even if I presume that I know them all, there is still an issue regarding the feasibility. So, the choice of applications might not be unanimous, for want of a better word.

In fine, I must say that I have enjoyed writing this book. I hope readers will equally enjoy reading it. They are very much encouraged to point out possible errors, typos, and suggestions to my e-mail address: mhmortad@gmail.com.

Oran, Algeria   Mohammed Hichem Mortad
July 5, 2022

Contents

Contents

Chapter 1
Classical Versions and Some Historical Notes

1.1 Finite-Dimensional Versions

It all started in 1942 when the legendary J. von Neumann established the following result:

Theorem 1.1.1 ([23]) *Let A and B be two square matrices of order n. Then*

$$\|AB - BA\|_F^2 - \|AB^* - B^*A\|_F^2 = -\operatorname{tr}[(A^*A - AA^*)(B^*B - BB^*)].$$

where $\|\cdot\|_F$ is the Frobenius norm, i.e., $\|A\|_F = \sqrt{\operatorname{tr}(A^*A)}$.
In particular, if either A or B is normal, then

$$AB = BA \iff AB^* = B^*A.$$

Proof Using basic properties of the trace of matrices, we easily obtain:

$$
\begin{aligned}
\|AB - BA\|_F^2 &= \operatorname{tr}[(AB - BA)^*(AB - BA)] \\
&= \operatorname{tr}(B^*A^*AB - B^*A^*BA - A^*B^*AB + A^*B^*BA) \\
&= \operatorname{tr}(B^*A^*AB) - \operatorname{tr}(B^*A^*BA) - \operatorname{tr}(A^*B^*AB) + \operatorname{tr}(A^*B^*BA) \\
&= \operatorname{tr}(A^*ABB^*) - \operatorname{tr}(AB^*A^*B) - \operatorname{tr}(BA^*B^*A) + \operatorname{tr}(AA^*B^*B) \\
&= \operatorname{tr}(A^*ABB^*) - \operatorname{tr}(AB^*A^*B) - \operatorname{tr}(ABA^*B^*) + \operatorname{tr}(AA^*B^*B) \\
&= \operatorname{tr}(A^*ABB^* - AB^*A^*B - ABA^*B^* + AA^*B^*B).
\end{aligned}
$$

By replacing each B by B^*, we get

$$\|AB^* - B^*A\|_F^2 = \operatorname{tr}(A^*AB^*B - ABA^*B^* - AB^*A^*B + AA^*BB^*).$$

M. H. Mortad, *The Fuglede-Putnam Theory*, Lecture Notes in Mathematics 2322,
https://doi.org/10.1007/978-3-031-17782-8_1

Hence

$$\|AB - BA\|_F^2 - \|AB^* - B^*A\|_F^2$$
$$= \mathrm{tr}(-A^*AB^*B + A^*ABB^* + AA^*B^*B - AA^*BB^*)$$
$$= \mathrm{tr}[-(A^*A - AA^*)(B^*B - BB^*)]$$
$$= -\mathrm{tr}[(A^*A - AA^*)(B^*B - BB^*)],$$

as desired.

So, if B (or A) is normal, then the previous equation implies that

$$\|AB - BA\|_F = \|AB^* - B^*A\|_F,$$

from which we derive

$$AB = BA \iff AB^* = B^*A,$$

as wished. $\square$

What is interesting about J. von Neumann's proof is the fact that it does not require the complex structure of the field. Simpler proofs on complex spaces are also known. For example:

Theorem 1.1.2 *Let A and B be two square matrices of the same size, such that, e.g., A is normal. Then*

$$AB = BA \iff A^*B = BA^*.$$

Proof *([35])* First, assume that A is a diagonal matrix where its diagonal elements are λ_i, with $i = 1, \cdots, n$. Since $AB = BA$, by taking $B = (b_{i,j})$ it follows that $\lambda_i b_{i,j} = b_{i,j}\lambda_j$ for all i and all j. That is, $(\lambda_i - \lambda_j)b_{i,j} = 0$. Hence $\overline{(\lambda_i - \lambda_j)}b_{i,j} = 0$. Therefore, $\overline{\lambda_i}b_{i,j} = \overline{\lambda_j}b_{i,j}$. In other words, $A^*B = BA^*$.

For the general case, we know that $A = U^*DU$ for some unitary matrix U where D is a diagonal matrix. Hence $AB = BA$ becomes $U^*DUB = BU^*DU$ or $D(UBU^*) = (UBU^*)D$. By the first part of the proof, we have

$$D^*(UBU^*) = (UBU^*)D^*.$$

Left multiplying by U^* and right multiplying by U, we obtain $U^*D^*UB = BU^*D^*U$, i.e., $A^*B = BA^*$.

To show the other implication, just observe that A^* is normal, then apply the previous proof to obtain from $A^*B = BA^*$ that $A^{**}B = BA^{**}$ or $AB = BA$, as needed. $\square$

Remark Another proof appeared in [13] (cf. [10]).

Remark S.K. Berberian obtained in [2] a Fuglede-Putnam theorem for finite-dimensional quaternionic Hilbert space (cf. [1]).

So much for proofs in a finite-dimensional setting.

1.2 The Classical Fuglede-Putnam Theorem

J. von Neumann wondered, in [23], whether his result may be established in the more general case of bounded linear operators on an infinite-dimensional Hilbert space? B. Fuglede was the first mathematician to answer this problem affirmatively in [11], not just for bounded operators infinite-dimensional spaces, but downright for unbounded operators. It is important to tell readers that P.R. Halmos obtained in [15], and almost simultaneously as B. Fuglede, a quite distinct proof of the theorem. More precisely, at the end of August 1949, B. Fuglede communicated his proof to P.R. Halmos at the Boulder meeting of the American Mathematical Society. P.R. Halmos' proof mainly dealt with the all bounded version, however, P.R. Halmos indicated that only minor modifications were needed to adapt his proof to the more general situation of unbounded operators.

B. Fuglede's version is the following:

Theorem 1.2.1 ([11]) *If* $T \in B(H)$, *and if* A *is normal and not necessarily bounded, then*

$$TA \subset AT \iff TA^* \subset A^*T.$$

Remark By the previous theorem, $BA = AB$ iff $BA^* = A^*B$ whenever $A \in B(H)$ is normal and $B \in B(H)$. So, we may then just drop the assumption $BA^* = A^*B$ in Theorem A.1.1.

Put differently, this says that if $TA \subset AT$, where A is normal, then $TE(\Delta) = E(\Delta)T$ for any Borel set Δ, where E is the spectral measure associated with A. This constitutes the first big application of Theorem 1.2.1.

The first generalization of Theorem 1.2.1 is due to C.R. Putnam in [25].

Theorem 1.2.2 ([25]) *Let* $T \in B(H)$ *and let* A *and* B *be two normal but non-necessarily bounded operators. Then*

$$TA \subset BT \iff TA^* \subset B^*T.$$

Hence $TE(\Delta) = F(\Delta)T$ *for any Borel set* Δ, *where* E *and* F *are the spectral measures associated with* A *and* B *respectively.*

Remark Let us all agree to call the latter version "the Fuglede-Putnam theorem" (even if we use it for A and $B = A^*$, where A is normal), and the former one as the "Fuglede theorem".

There is a particular terminology for the transformation that occurs in the Fuglede-Putnam theorem.

Definition 1.2.1 Say that $T \in B(H)$ intertwines two operators A, B when $TA \subset BT$, with "$\subset$" replaced by "=" if also $A, B \in B(H)$. We call the operator T the "intertwiner" of A and B.

Now, we give a proof of the Fuglede-Putnam theorem due to M. Rosenblum [30]:

Proof First, we deal with the bounded case, that is, we assume that all $T, A, B \in B(H)$.

Since $TA = BT$, we may show by induction that

$$TA^n = B^n T \text{ for } n = 0, 1, 2, \cdots,$$

thereby

$$Tp(A) = p(B)T,$$

where $p(z)$ is a certain polynomial in z of degree n, say. By the functional calculus for normal operators, we know that $e^{i\bar{z}A}$ and $e^{i\bar{z}B}$ are limits of polynomials in A and B respectively. Therefore

$$Te^{i\bar{z}A} = e^{i\bar{z}B}T, \; \forall z \in \mathbb{C}.$$

Set $f(z) = e^{-izB^*}Te^{izA^*}$. Then f is a function from $\mathbb{C}$ into $B(H)$. By the normality of A and B, we know that

$$e^A e^{A^*} = e^{A+A^*} \text{ and } e^B e^{B^*} = e^{B+B^*}.$$

Since $e^{-i\bar{z}B}Te^{i\bar{z}A} = T$, we may write

$$\begin{aligned} f(z) &= e^{-izB^*}Te^{izA^*} \\ &= e^{-izB^*}e^{-i\bar{z}B}Te^{i\bar{z}A}e^{izA^*} \\ &= e^{-izB^*-i\bar{z}B}Te^{i\bar{z}A+izA^*} \\ &= e^{-i(zB^*+\bar{z}B)}Te^{i(zA^*+\bar{z}A)}. \end{aligned}$$

The self-adjointness of both $zB^* + \bar{z}B$ and $zA^* + \bar{z}A$ yields the unitarity of both $e^{-i(zB^*+\bar{z}B)}$ and $e^{i(zA^*+\bar{z}A)}$ respectively. Hence

$$\|f(z)\| = \|e^{-i(zB^*+\bar{z}B)}Te^{-i(zA^*+\bar{z}A)}\| = \|T\|,$$

whichever $z \in \mathbb{C}$. Therefore, f is bounded, and since it is entire, Liouville's theorem implies that f is constant. Therefore, $f'(z) = 0$ for all $z \in \mathbb{C}$. In other words,

$$0 = f'(z) = -iB^*e^{-izB^*}Te^{izA^*} + ie^{-izB^*}TA^*e^{izA^*}.$$

Taking $z = 0$ yields

$$0 = -iB^*T + iTA^* \text{ or } TA^* = B^*T,$$

as coveted.

Assume now that A and B are unbounded normal operators. First, observe that $D(TA) \subset D(BT)$ does imply $D(TA^*) \subset D(B^*T)$. Indeed, $D(TA) = D(A) = D(A^*) = D(TA^*)$, and so $D(TA^*) \subset D(BT)$. However, $D(BT) = D(B^*T)$ for $D(B) = D(B^*)$ as B is normal.

Let m and n be both in $\mathbb{N}$, then consider the closed balls

$$B_n = \{z \in \mathbb{C} : |z| \le n\} \text{ and } B_m = \{z \in \mathbb{C} : |z| \le m\}.$$

If $P_{B_n}(A)$ and $P_{B_m}(B)$ are the spectral projections associated with A and B respectively, then $A_n := AP_{B_n}(A)$ and $B_m := BP_{B_m}(B)$ are bounded normal operators.

As $TA \subset BT$, then TA and BT coincide on $D(TA) = D(A)$. Since $\operatorname{ran} P_{B_n}(A) \subset D(A)$, $TAP_{B_n}(A) = BTP_{B_n}(A)$. Left multiplying by $P_{B_m}(B)$, then right multiplying by $P_{B_n}(A)$, as well as using some basic properties of the spectral measure all yield:

$$[P_{B_m}(B)TP_{B_n}(A)]A_n = B_m[P_{B_m}(B)TP_{B_n}(A)].$$

Since $P_{B_m}(B)TP_{B_n}(A)$ is bounded and everywhere defined, and A_n and B_m are bounded normal operators, we see that we are in a position to use the first part of the proof to obtain

$$[P_{B_m}(B)TP_{B_n}(A)](AP_{B_n}(A))^* = (BP_{B_m}(B))^*[P_{B_m}(B)TP_{B_n}(A)]$$

or

$$[P_{B_m}(B)TP_{B_n}(A)]P_{B_n}(A)A^* = B^*P_{B_m}(B)[P_{B_m}(B)TP_{B_n}(A)]$$

Thus,

$$P_{B_m}(B)TA^*P_{B_n}(A)x = P_{B_m}(B)B^*TP_{B_n}(A)x, \; \forall x \in D(A) = D(A^*)$$

(and hence $Tx \in D(B) = D(B^*)$). By sending both n and m to ∞, we get $P_{B_n}(A) \to I$ and $P_{B_m}(B) \to I$ respectively (both limits are with respect to the strong operator topology). Consequently,

$$TA^*x = B^*Tx, \ \forall x \in D(A) \ (\subset D(B^*T) = D(A)).$$

Thus $TA^* \subset B^*T$.

The other implication is now clear. Indeed, once we have shown the implication "$\Rightarrow$", assuming that $TA^* \subset B^*T$ gives $TA^{**} \subset B^{**}T$ for A^* and B^* are normal. In other words, $TA \subset BT$ since normal operators are already closed. The proof of the theorem is thus complete.   □

Proposition 1.2.3 *Let* $T \in B(H)$ *and let* A *and* B *be two normal but non-necessarily bounded operators. Then*

$$TA = BT \iff TA^* = B^*T.$$

Proof To show that, e.g., $TA = BT \Rightarrow TA^* = B^*T$, it suffices to check that $D(TA) = D(BT)$ gives $D(TA^*) = D(B^*T)$. Since $T \in B(H)$ and both A and B are normal, we have

$$D(TA^*) = D(A^*) = D(A) = D(TA) = D(BT) = D(B^*T),$$

as needed.   □

Remark There are different proofs of the Fuglede-Putnam theorem beside those mentioned above. Perhaps, the most elegant and the most known proof is the one due to M. Rosenblum [30], given just above. Unfortunately, an equally elegant proof (in the bounded case), unknown even to many specialists, is due to C.R. Putnam himself in [26]. This will be seen below. Other proofs may be consulted in [14, 24, 27, 28], and [29] (cf. [18] and [19]).

The following consequence is easily obtained:

Proposition 1.2.4 *Let* A *be a non-necessarily bounded normal operator and let* $B \in B(H)$. *Then*

$$BA \subset AB \iff BA^* \subset A^*B \iff B^*A \subset AB^* \iff B^*A^* \subset A^*B^*.$$

Proof

$$BA \subset AB \Longrightarrow BA^* \subset A^*B \text{ (Fuglede's theorem applied to} A)$$
$$\Longrightarrow (BA^*)^* \supset (A^*B)^*$$
$$\Longrightarrow B^*A \subset AB^* \text{ (for } A \text{ is closed)}$$
$$\Longrightarrow B^*A^* \subset A^*B^* \text{ (Fuglede's theorem applied to } A)$$

$$\Longrightarrow (B^*A^*)^* \supset (A^*B^*)^*$$

$$\Longrightarrow BA \subset AB.$$

$\square$

A simple argument allows us to drop the normality of A in the Fuglede theorem. Indeed, we have the same outcome by assuming that only $\overline{A}$ is normal (recall in passing that the normality of $\overline{A}$ does not yield in general that of A, see, e.g., [22]).

Lemma 1.2.5 *Let $B \in B(H)$ and let A be a densely defined operator such that $\overline{A}$ is normal. If $BA \subset AB$, then*

$$BA^* \subset A^*B \text{ and } B^*\overline{A} \subset \overline{A}B^*.$$

Proof *([3])* Since $\overline{A}$ is normal, $\overline{A}^* = A^*$ stays normal. Now,

$$BA \subset AB \Longrightarrow B^*A^* \subset A^*B^* \text{ (by taking adjoints)}$$

$$\Longrightarrow B^*\overline{A} \subset \overline{A}B^* \text{ (by Fuglede's theorem)}$$

$$\Longrightarrow BA^* \subset A^*B \text{ (by taking adjoints)}.$$

This marks the end of the proof. $\square$

Remark It is worth noticing that it is claimed in Lemma 2.8 of [32] that: *If $B \in B(H)$ and T is densely defined such that $\overline{T}$ exists, then*

$$BT \subset TB \Longleftrightarrow B\overline{T} \subset \overline{T}B.$$

However, and as observed in [22], only the implication "$\Longrightarrow$" holds. Let us give here as well the counterexample that appeared in [22] for ease of reference (there might be an even simpler example) :

Example 1.2.6 Let $H = L^2(\mathbb{R})$ and let $B = \mathcal{F}$ be the usual $L^2(\mathbb{R})$-Fourier transform. Consider also $T = I_{C_0^\infty}$, that is, the identity operator restricted to $C_0^\infty(\mathbb{R}) \subset L^2(\mathbb{R})$. Then $\overline{T} = I$, i.e., the full identity operator on $L^2(\mathbb{R})$.

Now, clearly $B\overline{T} \subset \overline{T}B$ (in fact, $B\overline{T} = \overline{T}B$). However, $BT \not\subset TB$ for if $BT \subset TB$ were true, we would have

$$D(BT) = D(T) = C_0^\infty(\mathbb{R}) \subset D(TB) = \{f \in L^2(\mathbb{R}) : \mathcal{F}f \in C_0^\infty(\mathbb{R})\},$$

which is evidently untrue. Indeed, let f be any *nonzero* function in $C_0^\infty(\mathbb{R})$. Then $f \in L^2(\mathbb{R})$ but $\mathcal{F}f \notin C_0^\infty(\mathbb{R})$ (why?), i.e., $f \notin D(TB)$, as needed.

Using a similar argument as the one above or a Berberian-like idea easily yields (cf. [9]):

Theorem 1.2.7 *Let $B \in B(H)$ and let N, M be densely defined closable operators such that $\overline{N}$ and $\overline{M}$ are normal. If $BN \subset MB$, then*

$$BN^* \subset M^*B.$$

We now come to a primordial observation. Obviously, Putnam's version generalizes Fuglede's. What is a priori unexpected, and also remarkable, is the fact that Fuglede's version does yield Putnam's, too. This extremely interesting finding is due to S.K. Berberian in [2], and it will henceforth be referred to as "Berberian's trick".

Proposition 1.2.8 *Fuglede's version implies Putnam's.*

Proof Let T, A, $B \in B(H)$ and let A and B be normal. Assume that $TA = BT$. We need to show that $TA^* = B^*T$ using Fuglede's version only. Take

$$\tilde{A} = \begin{pmatrix} B & 0 \\ 0 & A \end{pmatrix} \text{ and } \tilde{T} = \begin{pmatrix} 0 & T \\ 0 & 0 \end{pmatrix},$$

that are two matrices of operators defined on $H \oplus H$, where 0 designates the zero operator on H. Hence

$$\tilde{A}^* = \begin{pmatrix} B^* & 0^* \\ 0^* & A^* \end{pmatrix} = \begin{pmatrix} B^* & 0 \\ 0 & A^* \end{pmatrix}.$$

Then we have

$$\tilde{A}^*\tilde{A} = \begin{pmatrix} B^*B & 0 \\ 0 & A^*A \end{pmatrix} = \begin{pmatrix} BB^* & 0 \\ 0 & AA^* \end{pmatrix} = \tilde{A}\tilde{A}^*$$

because A and B are normal and therefore $\tilde{A}$ too is normal.

Furthermore, and as $TA = BT$, we can easily check that $\tilde{T}\tilde{A} = \tilde{A}\tilde{T}$. Applying the Fuglede theorem to the previous equation yields $\tilde{T}\tilde{A}^* = \tilde{A}^*\tilde{T}$, i.e.,

$$\begin{pmatrix} 0 & TA^* \\ 0 & 0 \end{pmatrix} = \begin{pmatrix} 0 & B^*T \\ 0 & 0 \end{pmatrix}.$$

Accordingly, $TA^* = B^*T$, as required.

The proof for unbounded A and B is identical, hence omitted.   □

Remark Readers will have other opportunities throughout this manuscript to see how to apply Berberian's trick for other proofs in the unbounded case.

As mentioned earlier, we give Putnam's much simpler proof of the Fuglede-Putnam theorem obtained over three decades after [25]. As already noted, this proof is almost unknown to many Fuglede-Putnam theorists. To prove it, C.R. Putnam [26] showed an auxiliary result which, according to the referee of his paper, was essentially contained in Theorem 6 and its proof in [4]. A somewhat modified proof was given by C.R. Putnam. We state this theorem without proof.

Theorem 1.2.9 ([26]) *Let H be a complex separable Hilbert space. Let $\{N_1, N_2, \cdots\}$ be a uniformly bounded ($\|N_k\| < const < \infty$), countable set of normal operators on H with the spectral resolutions*

$$N_k = \int z\, dE_z^k, \quad k = 1, 2, \cdots.$$

Then there exists a sequence of polynomials $\{p_1(z), p_2(z), \cdots\}$ and a sequence of closed sets $\{\alpha_1, \alpha_2, \cdots\}$ in the complex plane for which

$$E^k(\alpha_n) \longrightarrow I \text{ and } p_n(N_k)E^k(\alpha_n) \longrightarrow N_k^* \text{ (strongly), as } n \longrightarrow \infty,$$

for each fixed $k = 1, 2, \cdots$.

Remarks

1. The polynomials $p_n(z)$ and the closed sets α_n do not depend on k.
2. The above result only holds for bounded operators.

As an immediate consequence, we have an extremely short and elegant proof alike of the Fuglede-Putnam theorem.

Corollary 1.2.10 *Let $A, N_1, N_2 \in B(H)$ be such that N_1 and N_2 are normal. If $AN_1 = N_2 A$, then $AN_1^* = N_2^* A$.*

Proof ([26]) Obviously, $Ap(N_1) = p(N_2)A$ for any polynomial $p(z)$. So, if N_1 and N_2 have spectral resolutions as in Theorem 1.2.9, and if we choose $\{p_n(z)\}$ and $\{\alpha_n\}$ as in Theorem 1.2.9, then

$$E^2(\alpha_n)Ap_n(N_1)E^1(\alpha_n) = E^2(\alpha_n)p_n(N_2)AE^1(\alpha_n),$$

and so with respect to the strong operator topology:

$$AN_1^* = \lim_{n\to\infty} E^2(\alpha_n)Ap_n(N_1)E^1(\alpha_n) = \lim_{n\to\infty} E^2(\alpha_n)p_n(N_2)AE^1(\alpha_n) = N_2^*A,$$

and that's all! □

1.3   Weiss' Theorem Et al.

G. Weiss [33] obtained an interesting generalization of the Fuglede-Putnam theorem involving four normal operators. Let us state it in a slightly simplified form (see Theorem 2.2.11 for the full version).

Theorem 1.3.1 ([33]) *Let $A, N_1, N_2, M_1, M_2 \in B(H)$, where M_1, M_2, N_1 and N_2 are all normal. Assume further that $N_1 N_2 = N_2 N_1$ and that $M_1 M_2 = M_2 M_1$. Then*

$$N_1 A M_1 = N_2 A M_2 \implies N_1^* A M_1^* = N_2^* A M_2^*.$$

Remark By taking $N_1 = M_2 = I$, we recover the classical version of Fuglede-Putnam.

Remark The paper [5] contains a generalization of Theorem 1.3.1.

G. Weiss' original proof was fairly long. However, C.R. Putnam was able, once again, to give a greatly simple proof by falling back on Theorem 1.2.9.

Proof *([26])* For $k = 1$ and 2, let $\{E_z^k\}$ and $\{F_z^k\}$ denote the spectral families of N_k and M_k respectively. By the Fugelde theorem, we know that the families $\{E_z^1\}$ and $\{E_z^2\}$ commute, as N_1 commutes with N_2. Similarly, $\{F_z^1\}$ and $\{F_z^2\}$ commute. Since $N_1 A M_1 = N_2 A M_2$, it follows that

$$N_1^2 A M_1^2 = N_1 N_2 A M_2 M_1 = N_2 N_1 A M_1 M_2 = N_2^2 A M_2^2.$$

Hence, we know how to get $p(N_1) A p(M_1) = p(N_2) A p(M_2)$ for any polynomial $p(z)$. Choosing $\{p_n(z)\}$ and $\{\alpha_n\}$ as in Theorem 1.2.9 yields (strongly)

$$
\begin{aligned}
N_1^* A M_1^* &= \lim_{n \to \infty} E^2(\alpha_n) E^1(\alpha_n) p_n(N_1) A p_n(M_1) F^1(\alpha_n) F^2(\alpha_n) \\
&= \lim_{n \to \infty} E^1(\alpha_n) E^2(\alpha_n) p_n(N_2) A p_n(M_2) F^2(\alpha_n) F^1(\alpha_n) \\
&= N_2^* A M_2^*,
\end{aligned}
$$

and the proof is complete.   □

In the end, we state some spectacular generalization of the Fuglede-Putnam theorem (a proof may be consulted in [7]):

Theorem 1.3.2 ([7]) *Let $N = \begin{pmatrix} A & B \\ C & D \end{pmatrix}$ be normal on $H \oplus K$, where H and K are two Hilbert spaces. Let $T, S \in B(H, K)$ satisfy the equation*

$$XBX + XA - DX - C = 0,$$

where $I + T^*S$ is invertible. Then T satisfies the equation

$$XC^*X + XA^* - D^*X - B^* = 0$$

if and only if S does.

Remark To see why this is a generalization of the Fuglede-Putnam theorem, it suffices to take $B = C = T = 0$.

Remark See [34] for another generalization.

1.4 Counterexamples

In this section, we give several counterexamples that show that some generalizations are not possible.

Neophytes should be wary that the Fuglede-Putnam theorem does not say that $TA = BT \Rightarrow T^*A = BT^*$, even under strong conditions. Here is a simple counterexample on a bi-dimensional space:

Example 1.4.1 ([31]) Let

$$T = \begin{pmatrix} 1 & 1 \\ -1 & 1 \end{pmatrix}, \ A = \begin{pmatrix} 0 & 1 \\ 1 & 0 \end{pmatrix} \text{ and } B = \begin{pmatrix} 1 & 0 \\ 0 & -1 \end{pmatrix}.$$

Then T, A and B are all normal (in fact, A and B are even self-adjoint and unitary). Now

$$TA = BT, \text{ whereas } T^*A \neq BT^*,$$

as it may be checked.

It is well-known that a normal $A \in B(H)$ satisfies $\ker A = \ker A^*$, and that the converse is not true. However, the condition $\ker A = \ker A^*$ could replace the normality of A in some situations (not many though). For example, see [8]. So, we ask whether $TA = AT$ implies $TA^* = A^*T$ when $A, T \in B(H)$ and $\ker A = \ker A^*$. The answer is negative. Indeed:

Example 1.4.2 Let A be an invertible operator, and so $\ker A = \ker A^* = \{0\}$. So if the above were true, then we would have $AA = AA \Rightarrow AA^* = A^*A$. This is obviously untrue as this would mean that an invertible operator is always normal.

Another possible generalization that comes to mind is whether $\|TN - MT\| = \|TN^* - M^*T\|$, where $T, N, M \in B(H)$ are such that N and M are normal? This claim is refuted by the following counterexample due to M. J. Crabb and P. G. Spain in [6] (note that this example had a different aim in their paper).

Example 1.4.3 On $\mathbb{C}^2$, let

$$T = \begin{pmatrix} 1+i & 2 \\ -1+i & -2 \end{pmatrix}, \; N = \begin{pmatrix} i & 0 \\ 0 & 0 \end{pmatrix} \text{ and } M = \begin{pmatrix} 1 & 0 \\ 0 & -1 \end{pmatrix}.$$

Then M and N are normal, and

$$TN - MT = \begin{pmatrix} -2 & -2 \\ -2 & -2 \end{pmatrix} \text{ and } TN^* - M^*T = \begin{pmatrix} -2i & -2 \\ 2i & -2 \end{pmatrix}.$$

Since $TN - MT$ is unitarily equivalent to the diagonal 2×2 matrix having 0 and -4 on its diagonal, $\|TN - MT\| = 4$. Since

$$(TN^* - M^*T)^*(TN^* - M^*T) = 8I,$$

$\|TN^* - M^*T\| = 2\sqrt{2}$.

Remarks

1. This counterexample kills all hope of other related generalizations. For example, neither $\|TN - MT\| \leq \|TN^* - M^*T\|$ nor $\|TN - MT\| \geq \|TN^* - M^*T\|$ needs to hold. The former is seen by the same example whilst the latter is obtained by changing N by N^* (M already being self-adjoint) in the above example.
2. Another possible generalization is $|TN - MT| = |TN^* - M^*T|$, where $|\cdot|$ denotes the usual modulus of an operator (cf. [21]). The example above is a counterexample here as well. Indeed, just remember that $\||S|\| = \|S\|$ where $S \in B(H)$.

Now, can we hope to have a four-operator version of the Fuglede-Putnam theorem? In other words, if M, N, A and B are in $B(H)$ such that N and M are normal and such that $AN = MB$, then does it follow that $AN^* = M^*B$? The answer is negative as seen next:

Example 1.4.4 ([20]) On $\mathbb{C}^2$, let

$$M = \begin{pmatrix} 0 & 1 \\ -1 & 0 \end{pmatrix}, \; N = \begin{pmatrix} 1 & 0 \\ 0 & -1 \end{pmatrix}, \; A = \begin{pmatrix} 0 & 1 \\ 1 & 0 \end{pmatrix}, \text{ and } B = \begin{pmatrix} -1 & 0 \\ 0 & -1 \end{pmatrix}.$$

As it is easily seen, all operators involved are unitary where, in addition, A and B are self-adjoint. Readers may also check that

$$AN = MB \text{ while } AN^* \neq M^*B.$$

In the above example, we may check that $BN^* = M^*A$. So, one might be tempted to believe that $AN = MB$ does entail $BN^* = M^*A$, which would remain a plausible generalization since $A = B$ gives the classical version. This is again untrue as seen below:

Example 1.4.5 ([22]) Let

$$A = \begin{pmatrix} 2 & 0 \\ 0 & 1 \end{pmatrix} \text{ and } B = \begin{pmatrix} 3 & 0 \\ 0 & 1 \end{pmatrix}$$

and put

$$N = A^{-1} = \begin{pmatrix} \frac{1}{2} & 0 \\ 0 & 1 \end{pmatrix} \text{ and } M = B^{-1} = \begin{pmatrix} \frac{1}{3} & 0 \\ 0 & 1 \end{pmatrix}.$$

Then A, B, N and M are all positive and invertible. Besides,

$$AN = MB = I.$$

Finally,

$$BN^* = \begin{pmatrix} \frac{3}{2} & * \\ * & * \end{pmatrix} \neq \begin{pmatrix} \frac{2}{3} & * \\ * & * \end{pmatrix} = M^*A.$$

Related to Lemma 1.2.5, we may ask if T is densely defined such that $\overline{T}$ is normal and $BT^* \subset T^*B$, where $B \in B(H)$, then $BT \subset TB$?

The answer is again negative.

Example 1.4.6 Consider B and T as in Example 1.2.6. Then $\overline{T}$ and T^* are both normal. Besides,

$$B\overline{T} = BT^* = T^*B = \overline{T}B \, (= B),$$

and yet $BT \not\subset TB$.

B. Fuglede asked in [11] whether $BA \subset AB$ leads to $B^*A \subset AB^*$, or equivalently $BA^* \subset A^*B$, when $B \in B(H)$ is normal, and A is closed and densely defined? This is a natural and interesting question, for it is true when $A \in B(H)$. B. Fuglede found a strong counterexample in [12] to be given below. Fuglede's counterexample is, however, not very well-known as it was apparently missed by some writers and referees alike (witness [16] and [17]).

Example 1.4.7 ([12]) Consider two closed symmetric operators T and T' such that $T \subset T'$. Consider an operator A_p on $\ell^2(\mathbb{Z})$ whose graph consists of pairs $(x, y) \in \ell^2(\mathbb{Z}) \times \ell^2(\mathbb{Z})$ for which (x_n, y_n) is in the graph of T' for every $n < p$, and in the graph of T for every $n \geq p$. Then A_p is closed and symmetric for every p.

Now, let $U(x_n) = (x_{n+1})$ be defined on $\ell^2(\mathbb{Z})$ (the bilateral shift). Then U is unitary and

$$U^* A_p U = A_{p+1} \supsetneq A_p.$$

Therefore, we have a unitary $U \in B(H)$, and a closed symmetric operator A such that $UA \subset AU$ and $A \neq U^* AU$. The latter just means that $U^* A \not\subset AU^*$, as needed.

Remark Another example may also be found in [12].

In the above example, A is closed and symmetric. Nevertheless, a counterexample is not available anymore when A is self-adjoint. This is seen next:

Proposition 1.4.8 *If* $U \in B(H)$ *is unitary and* A *is an unbounded self-adjoint operator such that* $UA \subset AU$, *then* $UA = AU$.

Proof Since $UA \subset AU$ and A is self-adjoint, upon passing to adjoints, we get $U^*A \subset AU^*$. Right and left multiplying by U then entail $AU \subset UA$, and so $AU = UA$, as needed. □

References

1. E.H. Benabdi, M. Barraa, The spectral theorem for quaternionic normal operators. Funct. Anal. arXiv:2006.05253
2. S.K. Berberian, Note on a theorem of Fuglede and Putnam. Proc. Am. Math. Soc. **10**, 175–182 (1959)
3. I. Boucif, S. Dehimi, M.H. Mortad, On the absolute value of unbounded operators. J. Oper. Theory **82/2**, 285–306 (2019)
4. J. Bram, Subnormal operators. Duke Math. J. **22**, 75–94 (1955)
5. M. Chō, V. Müller, Spectral commutativity of multioperators. Funct. Anal. Approx. Comput. **4/1**, 21–25 (2012)
6. M.J. Crabb, P.G. Spain, Commutators and normal operators. Glasgow Math. J. **18/2**, 197–198 (1977)
7. J. Daughtry, The quadratic Putnam-Fuglede theorem. Proc. Am. Math. Soc. **59/2**, 404–405 (1976)
8. S. Dehimi, M.H. Mortad, Right (or left) invertibility of bounded and unbounded operators and applications to the spectrum of products. Complex Anal. Oper. Theory **12/3**, 589–597 (2018)
9. S. Dehimi, M.H. Mortad, A. Bachir, Unbounded generalizations of the Fuglede-Putnam theorem. Rend. Istit. Mat. Univ. Trieste **54**(7), 9 (2022). https://doi.org/10.13137/2464-8728/33883
10. E. Deutsch, P.M. Gibson, H. Schneider, The Fuglede-Putnam theorem and normal products of matrices. Collection of articles dedicated to Olga Taussky Todd. Linear Algebra Appl. **13/1–2**, 53–58 (1976)
11. B. Fuglede, A commutativity theorem for normal operators. Proc. Natl. Acad. Sci. **36**, 35–40 (1950)
12. B. Fuglede, Solution to Problem 3. Math. Scand. **2**, 346–347 (1954)

13. S.R. Garcia, R.A. Horn, A second course in linear algebra, in *Cambridge Mathematical Textbooks* (Cambridge University Press, Cambridge, 2017)
14. F. Gesztesy, M. Malamud, M. Mitrea, S. Naboko, Generalized polar decompositions for closed operators in Hilbert spaces and some applications. Integr. Equ. Oper. Theory **64/1**, 83–113 (2009)
15. P.R. Halmos, Commutativity and spectral properties of normal operators. Acta Sci. Math. Szeged **12** (1950). Leopoldo Fejér Frederico Riesz LXX annos natis dedicata, Pars B, pp. 153–156
16. Z.J. Jabłoński, I.B. Jung, J. Stochel, Unbounded quasinormal operators revisited. Integr. Equ. Oper. Theory **79/1**, 135–149 (2014)
17. P.E.T. Jørgensen, Unbounded operators: perturbations and commutativity problems. J. Funct. Anal. **39/3**, 281–307 (1980)
18. M.H. Mortad, An application of the Putnam-Fuglede theorem to normal products of self-adjoint operators. Proc. Am. Math. Soc. **131/10**, 3135–3141 (2003)
19. M.H. Mortad, *Normal products of self-adjoint operators and self-adjointness of the perturbed wave operator on $L^2(\mathbb{R}^n)$*. Thesis (Ph.D.)-The University of Edinburgh (United Kingdom). ProQuest LLC, Ann Arbor, MI, 2003
20. M.H. Mortad, Yet more versions of the Fuglede-Putnam theorem. Glasgow Math. J. **51/3**, 473–480 (2009)
21. M.H. Mortad, On the absolute value of the product and the sum of linear operators. Rend. Circ. Mat. Palermo, II. Ser **68/2**, 247–257 (2019)
22. M.H. Mortad, *Counterexamples in Operator Theory* (Birkhäuser/Springer, Cham, 2022)
23. J. von Neumann, Approximative properties of matrices of high finite order. Portugaliae Math. **3**, 1–62 (1942)
24. F.C. Paliogiannis, On Fuglede's theorem for unbounded normal operators. Ricerche Mat. **51**, 261–264 (2002)
25. C.R. Putnam, On normal operators in Hilbert space. Am. J. Math. **73**, 357–362 (1951)
26. C.R. Putnam, Normal operators and strong limit approximations. Indiana Univ. Math. J. **32/3**, 377–379 (1983)
27. H. Radjavi, P. Rosenthal, Hyperinvariant subspaces for spectral and n-normal operators. Acta Sci. Math. (Szeged) **32**, 121–126 (1971)
28. H. Radjavi, P. Rosenthal, *Invariant Subspaces*, 2nd edn. (Dover Publications, Inc., Mineola, NY, 2003)
29. W. Rehder, On the adjoints of normal operators. Arch. Math. (Basel) **37/2**, 169–172 (1981)
30. M. Rosenblum, On a theorem of Fuglede and Putnam. J. Lond. Math. Soc. **33**, 376–377 (1958)
31. W. Rudin, *Functional Analysis*, 2nd edn. (McGraw-Hill, New York, 1991)
32. F. Tian, On commutativity of unbounded operators in Hilbert space. Thesis (Ph.D.) *The University of Iowa* (2011). http://ir.uiowa.edu/etd/1095/
33. G. Weiss, The Fuglede commutativity theorem modulo the Hilbert-Schmidt class and generating functions for matrix operators. II. J. Oper. Theory **5/1**, 3–16 (1981)
34. R. Whitley, Fuglede's commutativity theorem and $\bigcap R(T - \lambda)$. Canad. Math. Bull. **33/3**, 331–334 (1990)
35. F. Zhang, *Linear Algebra. Challenging Problems for Students*, 2nd edn. Johns Hopkins Studies in the Mathematical Sciences (Johns Hopkins University Press, Baltimore, MD, 2009)
36. R. Zeng, Young's inequality in compact operators-the case of equality. J. Inequal. Pure Appl. Math. **6/4**, 10 pp. (2005). Article 110

Chapter 2
Generalizations to Bounded Nonnormal Operators

One natural way of generalizing the Fuglede-Putnam theorem is to weaken the normality assumptions. This chapter is mainly devoted to such generalizations.

2.1 Barría's Lemma Et al.

The first result, which is due to J. Barría, can be incorporated into the Fuglede-Putnam versions, even though this was perhaps not the intention of J. Barría.

Proposition 2.1.1 ([10]) *Let V_1, $V_2 \in B(H)$ be two isometries such that $V_1^* V_2 = V_2 V_1^*$. Then $V_1 V_2 = V_2 V_1$.*

The previous result was then generalized in [14], and the proof is shorter as well (apparently, the writers of [14] missed J. Barría's paper).

Proposition 2.1.2 ([14]) *Let $A, B \in B(H)$ be such that $AB^* = B^*A$. If A commutes with B^*B and B commutes with A^*A, then $AB = BA$. In particular, if A and B are isometries, then*

$$AB^* = B^*A \Longrightarrow AB = BA.$$

Proof Since $AB^* = B^*A$, $BA^* = A^*B$. By making use of the assumptions, we may then write

$$(AB - BA)^*(AB - BA) = (B^*A^* - A^*B^*)(AB - BA)$$
$$= B^*A^*AB - B^*A^*BA - A^*B^*AB + A^*B^*BA$$
$$= 0.$$

Therefore $AB = BA$, as wished. □

© The Author(s), under exclusive license to Springer Nature Switzerland AG 2022
M. H. Mortad, *The Fuglede-Putnam Theory*, Lecture Notes in Mathematics 2322,
https://doi.org/10.1007/978-3-031-17782-8_2

As already discussed, another way of extending the Fuglede-Putnam theorem is to place a fourth operator in both the assumption and the conclusion. To the best of my knowledge, the first (and only?) paper that dealt with such a question is [39]. We have already seen some counterexamples that show the failure of such generalizations. So, is a generalization to four operators just a fantasy? Not quite as shown next:

Proposition 2.1.3 ([39], cf. Example 2.7.3) *Let M be an isometry and let N be a co-isometry. If A and B are such that $AN = MB$, then $BN^* = M^*A$.*

Proof We have

$$A = ANN^* = (AN)N^* = (MB)N^* = M(BN^*).$$

Then

$$M^*A = M^*MBN^* = BN^*.$$

□

Remark Observe that the operators A and B do not seem to be in the expected order in the conclusion, i.e., they should be the other way round. However, the outcome is still an acceptable generalization for setting $A = B$ takes us back to the known version.

Remark The previous corollary constitutes in some sense a generalization of J. Barría's lemma.

Corollary 2.1.4 *If M is an isometry and A is an operator such that $A^*M = M^*A^*$. Then $AM^* = MA$.*

2.2   Generalizations Modulo the Hilbert-Schmidt Class

The following generalization to nonnormal operators is due to S.K. Berberian, who then thought that his result was perhaps unexciting. I don't know what readers think of it nowadays, but I quite like it because it uses nifty arguments. S.K. Berberian should look at versions that have appeared since his paper; some of them, thankfully not all, are unexciting.

Theorem 2.2.1 ([12]) *Let $A, B, X \in B(H)$ be such that $AX = XB$. Assume further that X is in the Hilbert-Schmidt class. If either A and B^* are hyponormal **or** B is invertible and $\|A\|\|B^{-1}\| \leq 1$, then*

$$A^*X = XB^*.$$

To prove this theorem, recall that Hilbert-Schmidt class forms an ideal C_2 in the algebra $B(H)$, and that C_2 is a Hilbert space with respect to the inner product

$$(A, B) = \sum (Ae_i, Be_i) = \operatorname{tr}(B^*A) = \operatorname{tr}(AB^*),$$

where (e_i) is *any* orthonormal basis of H, and where "tr" denotes the trace of operators as defined in, e.g., [50]. More details and more properties that will be used may be consulted in [50] or [51] (see also [40]).

We also need two auxiliary lemmata.

Lemma 2.2.2 *Let* $T \in B(H)$ *be such that* $Tx = x$, *where* $x \in H$. *If* T *is a contraction, i.e.,* $\|T\| \le 1$; *or if* T *is hyponormal, then* $T^*x = x$.

Proof Let $Tx = x$. If T is a contraction, $\|T\| \le 1$ and so $\|T^*\| \le 1$. Hence

$$
\begin{aligned}
\|T^*x - x\|^2 &= \|T^*x\|^2 - \langle x, T^*x \rangle - \langle T^*x, x \rangle + \|x\|^2 \\
&= \|T^*x\|^2 - \langle Tx, x \rangle - \langle x, Tx \rangle + \|x\|^2 \\
&\le \|T^*\|^2 \|x\|^2 - 2\|x\|^2 + \|x\|^2 \\
&\le 0.
\end{aligned}
$$

Therefore, $T^*x = x$.

A similar argument works just as good when T is hyponormal. Indeed, the hyponormality of T signifies that $\|T^*x\| \le \|Tx\|$ for all $x \in H$. Whence

$$
\begin{aligned}
\|T^*x - x\|^2 &= \|T^*x\|^2 - \langle x, T^*x \rangle - \langle T^*x, x \rangle + \|x\|^2 \\
&\le \|Tx\|^2 - \langle Tx, x \rangle - \langle x, Tx \rangle + \|x\|^2 \\
&= \|x\|^2 - 2\|x\|^2 + \|x\|^2 = 0,
\end{aligned}
$$

and so $T^*x = x$, as wished. $\square$

Remark The first part of the previous lemma is due to B. Sz.-Nagy.

Lemma 2.2.3 *Let* $A, B \in B(H)$. *Define a linear operator* $\mathcal{T}$ *on* C_2 *by* $\mathcal{T}X = AXB$.

1. $\mathcal{T}$ *is bounded with* $\|\mathcal{T}\| \le \|A\| \|B\|$, *and its adjoint is given* $\mathcal{T}^*X = A^*XB^*$.
2. *If* A *and* B *are positive, then so is* $\mathcal{T}$.
3. *If* A *and* B^* *are hyponormal, then* $\mathcal{T}$ *is hyponormal.*

Proof We only give the proof of the last statement, which is borrowed from [12]. There, readers may also find proofs of the remaining statements as well as some other properties of $\mathcal{T}$.

By assumption, $AA^* \leq A^*A$ and $B^*B \leq BB^*$. We have to show that $\mathcal{T}\mathcal{T}^* \leq \mathcal{T}^*\mathcal{T}$. Clearly

$$(\mathcal{T}^*\mathcal{T} - \mathcal{T}\mathcal{T}^*)X = A^*AXBB^* - AA^*XB^*B$$
$$= (A^*A - AA^*)XBB^* + AA^*X(BB^* - B^*B),$$

from which we derive the hyponormality of $\mathcal{T}$.   □

We are therefore ready to show Theorem 2.2.1.

Proof

1. Let A and B^* be hyponormal: Assume that B is invertible, then consider the operator $\mathcal{T}Y = AYB^{-1}$ defined on C_2. By the hyponormality of both A and $(B^*)^{-1}$, the preceding lemma says that $\mathcal{T}$ is hyponormal. By assumption $AX = XB$, i.e., $\mathcal{T}X = X$, and so by invoking Lemma 2.2.2, we have $\mathcal{T}^*X = X$. In other words, $A^*X(B^{-1})^* = X$, whereby $A^*X = XB^*$.

   In case B is non-invertible, take a certain λ for which $B - \lambda$ is invertible, and apply the previous reasoning to the pair $A - \lambda$ and $B - \lambda$, instead of A and B respectively.

2. Let $\|A\| \|B^{-1}\| \leq 1$, and consider again $\mathcal{T}Y = AYB^{-1}$. Then $\|\mathcal{T}\| \leq 1$. Since by assumption $\mathcal{T}X = X$, it ensues by Lemma 2.2.2 that $\mathcal{T}^*X = X$, which again yields $A^*X = XB^*$, and this marks the end of the proof.

                                                                                    □

In [22], T. Furuta slightly improved Theorem 2.2.1 above. Before stating this result, recall that a k-quasihyponormal operator T is one that obeys $\|T^*T^k x\| \leq \|T^{k+1}x\|$ for all $x \in H$. Clearly, every hyponormal operator is k-quasihyponormal.

Proposition 2.2.4 *Let $A, B, X \in B(H)$ be such that $AX = XB$. Assume further that X is in the Hilbert-Schmidt class. If A is k-quasihyponormal and B^* is invertible and hyponormal, then*

$$A^*X = XB^*.$$

The proof basically resembles that of Theorem 2.2.1, and uses a lemma like Lemma 2.2.3, namely:

Lemma 2.2.5 *Let $A, B \in B(H)$. Define a linear operator $\mathcal{T}$ on C_2 by $\mathcal{T}X = AXB$. If A and B^* are k-quasihyponormal, then so is $\mathcal{T}$. If $\mathcal{T}X = X$, then $\mathcal{T}^*X = X$ too.*

Next, we prove Proposition 2.2.4:

Proof Consider the operator $\mathcal{T}Y = AYB^{-1}$ defined on C_2. Since $(B^*)^{-1}$ is hyponormal and A is k-quasihyponormal, we know that $\mathcal{T}$ too is k-quasihyponormal. By assumption, $\mathcal{T}X = X$, and so $\mathcal{T}^*X = X$, thereby $A^*X = XB^*$, as required.   □

G. Weiss obtained the following generalization:

Theorem 2.2.6 ([61], cf. [29] and [38]) *If $A, B \in B(H)$ are normal, then*

$$\|AX - XB\|_2 = \|A^*X - XB^*\|_2$$

for every $X \in B(H)$.

T. Furuta [23], and independently F. Kittaneh in [31], obtained the following result in which the normality was relaxed. Notice that this result generalizes Theorem 2.2.1.

Theorem 2.2.7 ([23]) *Let $A, B \in B(H)$ be such that A and B^* are hyponormal. Then*

$$\|AX - XB\|_2 \geq \|A^*X - XB^*\|_2$$

for every X in the Hilbert-Schmidt class, where $\| \cdot \|_2$ designates the Hilbert-Schmidt norm.

Proof Define a linear operator $\mathcal{T}$ on C_2 by $\mathcal{T}X = AX - XB$. Then it may be checked that $\mathcal{T}^*X = A^*X - XB^*$. Doing some arithmetic yields

$$(\mathcal{T}^*\mathcal{T} - \mathcal{T}\mathcal{T}^*)X = (A^*A - AA^*)X + X(BB^* - B^*B).$$

Hence $\mathcal{T}^*\mathcal{T} - \mathcal{T}\mathcal{T}^*$ is positive, and so $\mathcal{T}$ is hyponormal. In other words, $\|\mathcal{T}^*X\|_2 \leq \|\mathcal{T}X\|_2$, that is,

$$\|A^*X - XB^*\|_2 \leq \|AX - XB\|_2,$$

as required.   □

A simpler proof of the foregoing theorem, which was obtained by W. Gong in [26], is given next.

Proof

$$
\begin{aligned}
\|A^*X - XB^*\|_2 &= (A^*X, A^*X) - (XB^*, A^*X) - (A^*X, XB^*) + (XB^*, XB^*) \\
&= (AA^*X, X) - (AX, XB) - (XB, AX) + (XB^*B, X) \\
&\leq (A^*AX, X) - (AX, XB) - (XB, AX) + (XBB^*, X) \\
&= (AX, AX) - (AX, XB) - (XB, AX) + (XB, XB) \\
&= \|AX - XB\|_2,
\end{aligned}
$$

as looked forward to.   □

We may drop the "Hilbert-Schmidt hypothesis" on X, but at the cost of beefing up the assumptions on A and B. Put differently, we have:

Theorem 2.2.8 ([31]) *Let $A, B, X \in B(H)$ be such that A and B^* are subnormal. Then*

$$\|A^*X - XB^*\|_2 \leq \|AX - XB\|_2.$$

Proof Since A and B^* are subnormal, there is a Hilbert space K, and there are normal operators N and M, both defined on $H \oplus K$, that obey

$$N = \begin{pmatrix} A & R \\ 0 & C \end{pmatrix} \text{ and } M = \begin{pmatrix} B & 0 \\ S & D \end{pmatrix}.$$

Let $Y = \begin{pmatrix} X & 0 \\ 0 & 0 \end{pmatrix}$. So,

$$NY - YM = \begin{pmatrix} AX - XB & 0 \\ 0 & 0 \end{pmatrix}$$

and

$$N^*Y - YM^* = \begin{pmatrix} A^*X - XB^* & -XS^* \\ R^*X & 0 \end{pmatrix}.$$

In view of Theorem 2.2.6, we obtain $\|NY - YM\|_2 = \|N^*Y - YM^*\|_2$, i.e.,

$$\|AX - XB\|_2^2 = \|A^*X - XB^*\|_2^2 + \|XS^*\|_2^2 + \|R^*X\|_2^2.$$

Accordingly,

$$\|A^*X - XB^*\|_2 \leq \|AX - XB\|_2,$$

as wished.   □

Remark Given Theorem 2.2.7, it is natural to ask whether Theorem 2.2.8 is valid for hyponormal operators A and B^* (where $X \in B(H)$)? This was asked by T. Furuta in [23]. A partial answer was obtained by V. Shulman and L. Turowska in [53], where the writers established the generalization under the extra assumption that A is a hyponormal operator of finite multiplicity. Whether Shulman-Turowska's result holds without the latter hypothesis is unknown to me.

Remark Theorem 2.2.8 says in particular that if $A, B, X \in B(H)$ are such that A and B^* are subnormal, then $AX - XB \in C_2$ yields $A^*X - XB^* \in C_2$. F. Kittaneh provided in [31] an example of two subnormal operators A and B such that $AX - XB \in C_2$, and yet $A^*X - XB^* \notin C_2$.

Remark A. Abdessemed and E. B. Davies in [1] extended Weiss' theorem, that is, Theorem 2.2.6, to more general classes than C_2, namely to the Schatten-von Neumann ideals C_p, $1 \leq p < \infty$. A relevant paper is [52].

We still have generalizations of Weiss' theorem. We give the next result without proof (see [30] for a proof). Notice in passing that the proof uses a famous result by D. Voiculescu in [60] as well as Berberian's trick.

Theorem 2.2.9 ([32]) *Let* $N, M \in B(H)$ *be normal, and let* f *be a Lipschitz function on* $\sigma(N) \cup \sigma(M)$, *i.e.,* $|f(z) - f(w)| \leq k|z - w|$ *for some positive constant* k *and all* z, w *in* $\sigma(N) \cup \sigma(M)$. *Then*

$$\|f(N)X - Xf(M)\|_2 \leq k\|NX - XM\|_2$$

for all $X \in B(H)$, *where* k *is the same Lipschitz constant of the function* f.

Remark The foregoing result significantly improves one in [3]. See [28] for a generalization.

By taking $X = I$, one obtains an interesting inequality in some particular perturbation problems (see, e.g., [19, 20] and [21]).

Corollary 2.2.10 ([32]) *Let* $N, M \in B(H)$ *be normal, and let* f *be a Lipschitz function on* $\sigma(N) \cup \sigma(M)$, *i.e.,* $|f(z) - f(w)| \leq k|z - w|$ *for some positive constant* k *and all* z, w *in* $\sigma(N) \cup \sigma(M)$. *Then*

$$\|f(N) - f(M)\|_2 \leq k\|N - M\|_2.$$

Remark By setting $f(z) = |z|$, we get

$$\||N| - |M|\|_2 \leq \|N - M\|_2$$

for any normal $N, M \in B(H)$, hence generalizing a result that appeared in [4] (cf. [41]).

Remark V. Lauric has obtained an improvement of Theorem 2.2.9 to the case of subnormal N and M^*, with an extra condition on f. See [33]. He has also obtained a result for hyponormal N and M, with some more assumptions.

The remaining part of this section is devoted to the original version of Weiss' (other) generalization of the Fuglede-Putnam theorem. Remember that we have already provided an exceedingly simple proof of a slightly simpler version of it due to C. R. Putnam (this was Theorem 1.3.1).

Theorem 2.2.11 ([61]) *Let* $X, N_1, N_2, M_1, M_2 \in B(H)$, *where* M_1, M_2, N_1 *and* N_2 *are all normal. Assume further that* $N_1 N_2 = N_2 N_1$ *and that* $M_1 M_2 = M_2 M_1$. *Then*

$$\|N_1 X M_1 + N_2 X M_2\|_2 = \|N_1^* X M_1^* + N_2^* X M_2^*\|_2.$$

In particular, if one of these expressions is in C_2, *so is the other, and if one of them is 0, then the other is 0 too.*

T. Furuta provided a certain generalization, in which he again relaxed the normality at the cost of assuming that X is in the Hilbert-Schmidt class.

Theorem 2.2.12 ([24]) *Let* $A, B, C, D \in B(H)$. *Suppose that* A, B^*, C *and* D^* *are hyponormal and* $CA^* = A^*C$ *and* $BD^* = D^*B$. *Then*

$$\|A^* X D^* + C^* X B^*\|_2 \leq \|AXD + CXB\|_2$$

for every X *in the Hilbert-Schmidt class.*

Proof Define a linear operator $\mathcal{T}$ on C_2 by $\mathcal{T}X = AXD + CXB$. Then it may be checked that $\mathcal{T}^*X = A^*XD^* + C^*XB^*$, by a glance at Lemma 2.2.3, and a moment's thought.

By denoting the commutator $ST - TS$ by $[S, T]$, we may therefore write

$$(\mathcal{T}^*\mathcal{T} - \mathcal{T}\mathcal{T}^*)X = A^*(\mathcal{T}X)D^* + C^*(\mathcal{T}X)B^* - A(\mathcal{T}^*X)D - C(\mathcal{T}^*X)B$$

$$= A^*AXDD^* - AA^*XD^*D + C^*CXBB^* - CC^*XB^*B$$

$$+ A^*CXBD^* - AC^*XB^*D + C^*AXDB^* - CA^*XD^*B$$

$$= [A^*, A]XDD^* + AA^*X[D, D^*] + [C^*, C]XBB^*$$

$$+ CC^*X[B, B^*].$$

In fact, there are two remaining terms but they are both zero as $CA^* = A^*C$ and $BD^* = D^*B$.

By a glance at Lemma 2.2.3 and the hyponormality of each of A, B^*, C and D^*, we see that $\mathcal{T}^*\mathcal{T} - \mathcal{T}\mathcal{T}^*$ is positive or that $\mathcal{T}$ is hyponormal, i.e., $\|\mathcal{T}^*X\|_2 \leq \|\mathcal{T}X\|_2$ for all X. In other words, we have shown the inequality

$$\|A^* X D^* + C^* X B^*\|_2 \leq \|AXD + CXB\|_2$$

as desired.

$\square$

Corollary 2.2.13 *Let* $A, B, C, D \in B(H)$. *Assume that* A, B^*, C *and* D^* *are hyponormal and* $CA^* = A^*C$ *and* $BD^* = D^*B$. *If* X *is in the Hilbert-Schmidt class, then*

$$AXD = CXB \implies A^*XD^* = C^*XB^*.$$

2.3 Generalizations to Subnormal or Hyponormal Operators

T. Furuta [22] obtained the following generalization of the Fuglede-Putnam theorem:

Theorem 2.3.1 *Let* $A, B, X \in B(H)$ *be such that* $AX = XB$. *Assume that both* A *and* B^* *are subnormal. Then*

$$A^*X = XB^*.$$

Remark We have not included a proof of the previous theorem because its proof is a mere consequence of Theorem 2.2.8.

Now, we give the generalization of the Fuglede-Putnam theorem to hyponormal operators.

Theorem 2.3.2 ([37]) *Let* $A \in B(H)$ *be hyponormal and let* $X \in B(H)$. *Assume that* $XA^* = AX$. *Then* $XA = A^*X$.

Remark The choice of the assumption to be $XA^* = AX$ (where A is hyponormal) and not $XA = A^*X$ is not random. See Example 2.7.6 for a counterexample. Recall that when A is normal, then this question does not arise.

The proof of Theorem 2.3.2 seems to need the following lemma:

Lemma 2.3.3 ([11]) *If* T *is hyponormal, and* $\mathcal{M}$ *is invariant under* T, *then* $\mathcal{M}$ *reduces* T, *whenever* $T/\mathcal{M}$ *is normal.*

Let us now prove Theorem 2.3.2:

Proof *([37])* Write $X = S + iT$, where S and T are self-adjoint. Since $XA^* = AX$, $X^*A^* = AX^*$. Hence $SA^* = AS$, and also $TA^* = AT$.

Let $\mathcal{M} = \ker S$, then write $H = \mathcal{M}^\perp \oplus \mathcal{M}$. It is seen that $\mathcal{M}$ is invariant for A^*. By consulting Proposition 3.7 in [13] or else, we see that A and S may be expressed as:

$$A = \begin{pmatrix} C & D \\ 0 & E \end{pmatrix} \text{ and } S = \begin{pmatrix} R & 0 \\ 0 & 0 \end{pmatrix}.$$

Since A is hyponormal, it is seen that C too is hyponormal, but only on $\mathcal{M}^{\perp}$. Because $AS = SA^*$, we have $CR = RC^*$. Since R is one-to-one and has dense range, we obtain the normality of C by Theorem 3 in [48]. Hence $\mathcal{M}$ reduces A by say Lemma 2.3.3, thereby $D = 0$. Thus, A is reduced to the form $A = \begin{pmatrix} C & 0 \\ 0 & E \end{pmatrix}$. Since $CR = RC^*$, and C is normal, the classical Fuglede-Putnam theorem implies that $C^*R = RC$, which gives $A^*S = SA$. In a similar way, we may obtain $A^*T = TA$. Consequently, $XA = A^*X$, as suggested. $\qquad\square$

Remark The preceding theorem was, in fact, shown for the class of M-hyponormal operators. Recall that $A \in B(H)$ is said to be M-hyponormal if there exists a positive number M such that

$$\|(A - z)^*x\| \le M\|(A - z)x\|$$

for all $x \in H$ and all $z \in \mathbb{C}$.

By Berberian's trick, we may show:

Theorem 2.3.4 ([37]) *If $A, B, X \in B(H)$, where A and B are hyponormal, then*

$$XB^* = AX \Longrightarrow XB = A^*X.$$

Proof Let

$$\tilde{A} = \begin{pmatrix} A & 0 \\ 0 & B \end{pmatrix} \text{ and } \tilde{X} = \begin{pmatrix} 0 & X \\ 0 & 0 \end{pmatrix}$$

be defined on $H \times H$ where 0 stands for the zero operator on H. Hence

$$\tilde{A}\tilde{A}^* = \begin{pmatrix} AA^* & 0 \\ 0 & BB^* \end{pmatrix} \le \begin{pmatrix} A^*A & 0 \\ 0 & B^*B \end{pmatrix} = \tilde{A}^*\tilde{A}$$

for A and B are hyponormal, and hence $\tilde{A}$ is hyponormal too.

In addition, and as $XB^* = AX$, it is readily seen that $\tilde{X}\tilde{A}^* = \tilde{A}\tilde{X}$. Applying Theorem 2.3.2 to the previous equation gives $\tilde{X}\tilde{A} = \tilde{A}^*\tilde{X}$, i.e.,

$$\begin{pmatrix} 0 & XB \\ 0 & 0 \end{pmatrix} = \begin{pmatrix} 0 & A^*X \\ 0 & 0 \end{pmatrix}.$$

Thus, $XB = A^*X$, as wished. $\qquad\square$

Remark M. Radjabalipour obtained in [49] more general results than Theorem 2.3.4.

The next consequence was obtained in [31]:

Corollary 2.3.5 *Let $A, V, X \in B(H)$. If V is an isometry, A^* is hyponormal, and X is one-to-one, then $VX = XA$ implies that A is unitary.*

Proof Since an isometry is hyponormal, Theorem 2.3.4 gives $XA^* = V^*X$. Hence $V^*XA = X$ or $X = XA^*A$. By the "one-to-oness" of X, we get $A^*A = 1$. So, the hyponormal operator A^* is right invertible, and so it is invertible. Thus, A is unitary. □

Remark Using the same idea of proof as in [37], T. Yoshino provided an improvement of Theorems 2.3.2 and 2.3.4.

Before stating a useful lemma, recall that an operator $A \in B(H)$ is called dominant if there a real number M_λ for each λ such that

$$\|(A - \lambda)^*x\| \leq M_\lambda \|(A - \lambda)x\|$$

for all $x \in H$.

Lemma 2.3.6 ([48]) *If T is dominant, and M is invariant under T, then M reduces T, whenever T/M is normal.*

Theorem 2.3.7 ([62]) *If $A, B, C \in B(H)$ are such that $CA = BC$, where A^* is M-hyponormal and B is dominant, then $CA^* = B^*C$.*

Proof As above, by using Lemma 2.3.6. □

2.4 Generalizations to p-Hyponormal or log-Hyponormal Operators

First, recall that an operator $T \in B(H)$ is said to be p-hyponormal, where $p > 0$, if $(TT^*)^p \leq (T^*T)^p$. We say that T is log-hyponormal if T is invertible and $\log(TT^*) \leq \log(T^*T)$.

The next result will allow us to obtain a version of the Fuglede-Putnam theorem for p-hyponormal or log-hyponormal operators.

Theorem 2.4.1 ([59]) *Let $T \in B(H)$ be p-hyponormal or log-hyponormal and let $L \in B(H)$ be self-adjoint such that $TL = LT^*$. Then $T^*L = LT$.*

Remark The preceding result generalizes a version obtained by S. M. Patel in [47], who obtained it with some extra conditions.

Before giving the proof, we define the so-called Aluthge transform (introduced by A. Aluthge in [2]). Let $T = U|T|$ be the polar decomposition of $T \in B(H)$ in terms of partial isometries. The Aluthge transform is the operator $\widetilde{T} = $

$|T|^{1/2}U|T|^{1/2}$. Among the remarkable properties of the Aluthge transform, we have:

1. ([2, 56]) If T is p-hyponormal for $0 < p < 1$ (resp. log-hyponormal), then the Aluthge transform $\widetilde{T}$ is hyponormal if $p \geq 1/2$, and $(p + 1/2)$-hyponormal if $0 < p \leq 1/2$ (resp. $1/2$-hyponormal).
2. ([46]) A p-hyponormal operator whose Aluthge transform is normal, is itself normal.

Now, we prove Theorem 2.4.1.

Proof *([59])* We only give the proof of the theorem in the case T is p-hyponormal. The proof in the case of log-hyponormality may be consulted in [59].

The proof is somewhat technical and requires several auxiliary results. The aim is to get to a situation, where the Fuglede-Putnam theorem for normal operators is usable.

First, we show that $TL = LT^* = 0$ yields $T^*L = LT = 0$. Since $\ker T$ reduces T, $TL = 0$ gives $\operatorname{ran} L \subset \ker T \subset \ker T^*$. Also, $\overline{\operatorname{ran} T} \subset \ker L$. Therefore, $T^*L = LT = 0$.

Assume now that $TL \neq 0$. Write $H = \overline{\operatorname{ran} L} \oplus \ker L$ and

$$L = \begin{pmatrix} L_1 & 0 \\ 0 & 0 \end{pmatrix} \text{ and } T = \begin{pmatrix} T_1 & S \\ 0 & T_2 \end{pmatrix},$$

where L_1 is self-adjoint and one-to-one (hence L_1 has dense range), and T_1 is p-hyponormal (cf. [58]). Since $TL = LT^*$, it ensues that $T_1L_1 = L_1T_1^*$. Since $\ker T_1$ reduces T_1 and L_1, the latter may be expressed as $T_1 = T_{11} \oplus 0$ and $L_1 = L_{11} \oplus L_{22}$ with respect to the new decomposition $\overline{\operatorname{ran} L} = \overline{\operatorname{ran} |T_1|} \oplus \ker T_1$. Again, it may be checked that T_{11} is an injective p-hyponormal operator and that L_{11} is an injective self-adjoint operator obeying $T_{11}L_{11} = L_{11}T_{11}^*$.

Now, let $\widetilde{T_{11}}$ be the Aluthge transform of T_{11}. If $p \geq 1/2$, then it is easy to see that $\widetilde{T_{11}}W = W\widetilde{T_{11}}^*$, where $W = |T_{11}|^{1/2}L_{11}|T_{11}|^{1/2}$ is one-to-one and self-adjoint, and $\widetilde{T_{11}}$ is hyponormal. By Theorem 1 in [55], we know that $\widetilde{T_{11}}$ is normal, and so T_{11} is normal itself by Lemma 3 in [59]. Whence $T_1 = T_{11} \oplus 0$ too is normal. We are now in a position where wa can use the classical Fuglede-Putnam theorem. Hence $T_1^*L_1 = L_1T_1$. Since T_1 is normal, $S = 0$ by Lemma 12 in [59]. Thus, $T^*L = LT$.

Let us now obtain the same conclusion when $0 < p < 1/2$. In this case, $\widetilde{T_{11}}$ becomes an injective $(p + 1/2)$-hyponormal operator. Arguing as before, we may get that $\widetilde{T_{11}}$ is normal, whereby T_1 is normal. Accordingly, we obtain the same conclusion in the case $0 < p < 1/2$.   □

Proposition 2.4.2 ([59]) *Let $T \in B(H)$ be p-hyponormal or* log-*hyponormal and let $X \in B(H)$ be such that $TX = XT^*$. Then $T^*X = XT$.*

Proof Let $X = L + iJ$ be the Cartesian decomposition of X (hence L and J are self-adjoint). Since $TX = XT^*$, we have $X^*T^* = TX^*$. Hence $TL = LT^*$ and $TJ = JT^*$. By Theorem 2.4.1, $T^*L = LT$ and $T^*J = JT$. Thus

$$T^*X = T^*L + iT^*J = LT + iJT = XT,$$

as needed.   □

Thanks to the usual Berberian's trick, we have a more general Fuglede-Putnam version of the previous result.

Corollary 2.4.3 ([59]) *Let* $T^*, S \in B(H)$ *be p-hyponormal (resp. log-hyponormal) and let* $X \in B(H)$ *be such that* $SX = XT$. *Then* $S^*X = XT^*$.

Proof Put $A = \begin{pmatrix} T^* & 0 \\ 0 & S \end{pmatrix}$ and $B = \begin{pmatrix} 0 & 0 \\ X & 0 \end{pmatrix}$. Then A is p-hyponormal (resp. log-hyponormal). Besides, $BA^* = AB$, and so applying Proposition 2.4.2 followed by an examination of the entries of the matrices of operators lead to $XT^* = S^*X$.   □

2.5   A Generalization Based on the Julia Operator

Most of the material here is borrowed from [39]. Before giving the generalization, we wish to recall the following standard lemma.

Lemma 2.5.1 *Assume that N is unitary and that A and B are two bounded operators. Then*

$$NA = BN \implies NA^* = B^*N.$$

Now we wish to drop the "unitarity" hypothesis on N. We use a trick of matrix operators based on the Julia operator. Recall that if $A \in B(H)$ is a contraction, i.e., $\|A\| \le 1$, then the Julia operator is defined on $H \oplus H$ by:

$$J(A) = \begin{pmatrix} (I - |A^*|^2)^{\frac{1}{2}} & A \\ -A^* & (I - |A|^2)^{\frac{1}{2}} \end{pmatrix}.$$

A well-known result states that the Julia operator is unitary (see Exercise 6.3.18 in [40] for a detailed proof).

Theorem 2.5.2 (Cf. [44]) *Let A and B be two bounded operators. Suppose N is a contraction such that*

$$(1 - N^*N)^{\frac{1}{2}}A = B(1 - NN^*)^{\frac{1}{2}} = (1 - N^*N)^{\frac{1}{2}}A^* = B^*(1 - NN^*)^{\frac{1}{2}} = 0.$$

Then

$$NA = BN \implies NA^* = B^*N.$$

Proof Consider the matrix operators defined on $H \oplus H$ as

$$\tilde{A} = \begin{pmatrix} 0 & 0 \\ 0 & A \end{pmatrix}; \ \tilde{B} = \begin{pmatrix} B & 0 \\ 0 & 0 \end{pmatrix}$$

and

$$\tilde{N} = J(N) = \begin{pmatrix} (1 - NN^*)^{\frac{1}{2}} & N \\ -N^* & (1 - N^*N)^{\frac{1}{2}} \end{pmatrix}.$$

Then

$$\tilde{N}\tilde{A} = \begin{pmatrix} 0 & NA \\ 0 & (1 - N^*N)^{\frac{1}{2}}A \end{pmatrix} = \begin{pmatrix} 0 & NA \\ 0 & 0 \end{pmatrix}.$$

We also have

$$\tilde{B}\tilde{N} = \begin{pmatrix} B(1 - NN^*)^{\frac{1}{2}} & BN \\ 0 & 0 \end{pmatrix} = \begin{pmatrix} 0 & BN \\ 0 & 0 \end{pmatrix}.$$

Then $\tilde{B}\tilde{N} = \tilde{N}\tilde{A}$.

Since $\tilde{N}$ is unitary, the foregoing lemma yields

$$\tilde{B}^*\tilde{N} = \begin{pmatrix} B^*(1 - NN^*)^{\frac{1}{2}} & BN \\ 0 & 0 \end{pmatrix} = \begin{pmatrix} 0 & NA^* \\ 0 & (1 - N^*N)^{\frac{1}{2}}A^* \end{pmatrix} = \tilde{N}\tilde{A}^*.$$

The remaining unused two hypotheses allow us to get $B^*N = NA^*$, and this completes the proof.   □

Corollary 2.5.3 *Let A and B be two bounded operators. If N is an isometry such that*

$$B(1 - NN^*)^{\frac{1}{2}} = B^*(1 - NN^*)^{\frac{1}{2}} = 0,$$

then

$$BN = NA \implies B^*N = NA^*.$$

Remark The condition $B(1 - NN^*)^{\frac{1}{2}} = B^*(1 - NN^*)^{\frac{1}{2}} = 0$ cannot be completely eliminated in the previous corollary. For instance, consider the unilateral shift S on ℓ^2, which is an isometry, then set $N = B = S$. If we also set $A = S$, then

$$BN = S^2 = NA \text{ while } B^*N = S^*S \neq SS^* = NA^*.$$

We now give a non-trivial example satisfying the hypotheses of the previous corollary, and that is not satisfied by any other known versions of the Fuglede-Putnam theorem (see [39] for further details).

Example 2.5.4 Consider the infinite matrices

$$N = \begin{bmatrix} 0 & 0 & & & 0 \\ & 0 & 0 & 0 \\ & & 1 & 0 & \ddots & \ddots \\ & & 0 & 1 & 0 & \ddots \\ & & & 0 & 1 & \ddots & \ddots \\ 0 & & & & \ddots & \ddots \end{bmatrix}, \quad B = \begin{bmatrix} 0 & 0 & 0 & 0 & & & 0 \\ & 0 & 0 & 0 & 0 \\ & & 0 & 0 & 2 & 0 \\ & & & 0 & 0 & 1 & 0 \\ & & & & 0 & 0 & 1 & \ddots \\ 0 & & & & & \ddots & \ddots & \ddots \end{bmatrix}$$

and

$$A = \begin{bmatrix} 0 & 0 & 2 & 0 & & 0 \\ & 0 & 0 & 0 & 1 & 0 \\ & & 0 & 0 & 0 & 1 & \ddots \\ & & & 0 & 0 & 0 & \ddots \\ & & & & 0 & \ddots & \ddots \\ 0 & & & & & \ddots & \ddots \end{bmatrix}$$

Then $B(1 - NN^*)^{\frac{1}{2}} = B^*(1 - NN^*)^{\frac{1}{2}}$. Also, we have $BN = NA$ and hence $NA^* = B^*N$.

However, one can also check that B is not normal and neither is A. In fact, B is not hyponormal. It is not even dominant hence our operators are not covered by the Fuglede-Putnam versions recalled above.

2.6 Further Comments on Other Generalizations

There are still many more generalizations for other classes of non-normal operators. We will only list a few recent references. In the remaining of this section, set $XA = BX$, and so X is the intertwiner of A and B. The sought conclusion is $XA^* = B^*X$.

K. Tanahashi, S.M. Patel and A. Uchiyama established in [57] several Fuglede-Putnam theorems (p, q)-quasihyponormal operators, as did S. Mecheri in [34]. See also [9]. Recall that $T \in B(H)$ is said to be (p, q)-quasihyponormal if $T^{*k}[(T^*T)^p - (TT^*)^p]T^k \geq 0$, for some integer $k \geq 1$ and $0 < p \leq 1$.

A. Bachir and F. Lombarkia obtained in [6] generalizations for the class of w-hyponormal operators. Recall that $T \in B(H)$ is called w-hyponormal when $|\widetilde{T}^*| \leq |T| \leq |\widetilde{T}|$, where $\widetilde{T}$ designates the usual Aluthge transform. See also [5].

S. Mecheri, K. Tanahashi and A. Uchiyama established in [35] a Fuglede-Putnam version where either B is p-hyponormal and A^* is in the $\mathcal{Y}$ class or B^* is p-hyponormal and A is $\mathcal{Y}$ operator. See, e.g., [35] for the definition of the $\mathcal{Y}$ class of operators.

A. Bachir and T. Prasad obtained in [7] a generalization to the class of (α, β)-normal operators, where A is invertible and B is in the class of (α, β)-normal operators, where the intertwiner is in the Hilbert-Schmidt class. Recall in passing that following [15], say that $A \in B(H)$ is (α, β)-normal $(0 \leq \alpha \leq 1 \leq \beta)$ if:

$$\alpha^2 A^*A \leq AA^* \leq \beta^2 A^*A.$$

P. Pagacz found a paranormal operator S, a unitary U and an orthogonal projection such that $SP = PU$, but $S^*P \neq PU^*$. More detailed may be found in [45].

Other generalizations and related works, though not all the existing ones in the literature, are [8, 16–18, 25, 27, 36] and [43].

2.7 Counterexamples

We start by showing that the most naive generalization to nonnormal (even quasinormal) operators fails. That is, if $A, T \in B(H)$ are such that A is quasinormal and $TA = AT$, then it is not true that $TA^* = A^*T$.

Example 2.7.1 Let S be the shift operator on ℓ^2 (remember that $S^*S = I$). Then S is quasinormal. If the Fuglede theorem were true for quasinormal operators, then we would have

$$SS = SS \implies SS^* = S^*S = I,$$

that is, the shift would be unitary, and this is patently untrue.

Many results holding for normal operators A and B with $AB = BA$, do not necessarily remain valid if A and B are assumed to be in some class of nonnormal operators. The main reason is that in the event of normality we also have $AB^* = B^*A$ thanks to the Fuglede theorem. So, many authors when trying to generalize results that already hold for normal operators to nonnormal ones, assume both conditions $AB = BA$ and $AB^* = B^*A$. The next example shows that there are quasinormal operators that are not normal, yet they satisfy the latter equations.

Example 2.7.2 ([42]) Let $S \in B(\ell^2)$ be the usual shift operator, then define

$$A = \begin{pmatrix} S & 0 \\ 0 & 0 \end{pmatrix} \text{ and } B = \begin{pmatrix} 0 & 0 \\ 0 & S \end{pmatrix}.$$

on $\ell^2 \oplus \ell^2$. Since $S^*S = I$, clearly, $AA^*A = A^*AA$, and so A is quasinormal. Similarly, B is quasinormal. Neither A nor B is normal, and yet

$$AB = BA = AB^* = B^*A = 0_{\ell^2 \oplus \ell^2}.$$

Remark Readers might be interested in consulting [14], in which some situations guaranteeing that either A or B be normal are given, provided that A and B are hyponormal and obeying $AB = BA$ and $AB^* = B^*A$.

Let M be an isometry *or* let N be a co-isometry. Let A and B be two bounded linear operators such that $AN = MB$. Does it follow that $BN^* = M^*A$ (cf. Proposition 2.1.3)?

Example 2.7.3 ([39]) If one takes the shift operator S, defined on ℓ^2, then by setting $M = N = A = B = S$ (hence N is not a co-isometry), one sees that $AN = MB$ while $BN^* \neq M^*A$.

On the other hand, if one sets $M = N = A = B = S^*$ (hence M is not an isometry), then $AN = MB$ whereas $BN^* \neq M^*A$.

Readers are already aware that the Fuglede theorem need not hold for the class of quasinormal operators. Hence the Fuglede-Putnam theorem would not hold for quasinormal operators either. However, there is still a case that is worth investigating as regards the Fuglede-Putnam version, namely: Let $A, B, T \in B(H)$. Do we have "$TB = AT \Rightarrow TB^* = A^*T$", when A is normal and B is quasinormal? The answer is negative even when A is unitary.

Example 2.7.4 ([54]) Let A be the bilateral shift defined on $\ell^2(\mathbb{Z})$, i.e., a linear operator whose action on an orthonormal basis of $\ell^2(\mathbb{Z})$ is given by $Ae_n = e_{n+1}$, $n \in \mathbb{Z}$. Define B as

$$Be_n = \begin{cases} e_{n+1}, & n \geq 0, \\ 0, & n < 0 \end{cases}$$

and take $T = B$. Clearly A is unitary and B is quasinormal. Indeed, for all $n \in \mathbb{Z}$

$$B^* B B e_n = B B^* B e_n = \begin{cases} e_{n+1}, & n \geq 0, \\ 0, & n < 0. \end{cases}$$

Now, for all $n \in \mathbb{Z}$, we have

$$T B e_n = A T e_n = \begin{cases} e_{n+2}, & n \geq 0, \\ 0, & n < 0. \end{cases}$$

Thus, $T B = A T$. But

$$T B^* e_0 = 0 \neq e_0 = A^* T e_0,$$

and so $T B^* \neq A^* T$, as needed.

We have just seen above that if A and B are two quasinormal operators, then $AB = BA$ does not necessarily yield $AB^* = B^*A$. What about the converse question "$AB^* = B^*A \Rightarrow AB = BA$"? This seemingly innocent question is highly non-obvious.

The first example that appeared in the literature may be found in [14], and it was fairly long. Next, we give a simpler, also stronger, example (this appeared in [42]):

Example 2.7.5 Consider A and B

$$A e_n = \begin{cases} e_{n-1}, & n \leq 1, \\ 0, & n > 1. \end{cases} \quad \text{and} \quad B e_n = \begin{cases} e_{n+1}, & n < 1, \\ 0, & n \geq 1, \end{cases}$$

both defined on $\ell^2(\mathbb{Z})$, and where (e_n) is an orthonormal basis. Readers may readily check that

$$A^* e_n = \begin{cases} e_{n+1}, & n \leq 0, \\ 0, & n > 0. \end{cases} \quad \text{and} \quad B^* e_n = \begin{cases} e_{n-1}, & n \leq 1, \\ 0, & n > 1. \end{cases}$$

Hence

$$A^* A e_n = \begin{cases} e_n, & n \leq 1, \\ 0, & n > 1. \end{cases} \quad \text{and} \quad B^* B e_n = \begin{cases} e_n, & n \leq 0, \\ 0, & n > 0. \end{cases}$$

Therefore, it is seen that

$$A A^* A e_n = A^* A A e_n \text{ and } B B^* B e_n = B^* B B e_n,$$

both equations for all $n \in \mathbb{Z}$. This means that $AA^*A = A^*AA$ and $BB^*B = B^*BB$, that is, A and B are both quasinormal. Now, readers can also check that

$$AB^*e_n = e_{n-2} = B^*Ae_n, \ \forall n \le 1$$

and

$$AB^*e_n = 0 = B^*Ae_n, \ \forall n > 1.$$

Thus, $AB^* = B^*A$. Finally, since, e.g.,

$$ABe_1 = 0 \neq e_1 = BA(e_1),$$

we see that $AB \neq BA$, as wished.

Theorem 2.3.2 states that if $A \in B(H)$ is hyponormal and $XA^* = AX$, then $XA = A^*X$ for all $X \in B(H)$. Can a version like

$$XA = A^*X \Longrightarrow XA^* = AX$$

hold if A is hyponormal? To the best of my knowledge (and curiously!), no one has investigated this question. The answer is negative, even when A is quasinormal. The following counterexample is not documented anywhere, but a simpler example could be known to certain specialists.

Example 2.7.6 On $\ell^2(\mathbb{Z})$, consider a unitary operator A, a quasinormal operator B, and a T that obey $TB = AT$, but $TB^* \neq A^*T$ (as in Example 2.7.4). Set on $\ell^2(\mathbb{Z}) \oplus \ell^2(\mathbb{Z})$

$$X = \begin{pmatrix} 0 & T \\ 0 & 0 \end{pmatrix} \text{ and } \widetilde{A} = \begin{pmatrix} A^* & 0 \\ 0 & B \end{pmatrix}.$$

Then $\widetilde{A}$ is quasinormal. Moreover,

$$X\widetilde{A} = \begin{pmatrix} 0 & TB \\ 0 & 0 \end{pmatrix} = \begin{pmatrix} 0 & AT \\ 0 & 0 \end{pmatrix} = \widetilde{A}^*X.$$

Since $TB^* \neq A^*T$, it ensues that $X\widetilde{A}^* \neq \widetilde{A}X$, as wished.

References

1. A. Abdessemed, E.B. Davies, Some commutator estimates in the Schatten classes. J. Lond. Math. Soc. (2) **39/2**, 299–308 (1989)
2. A. Aluthge, On p-hyponormal operators for $0 < p < 1$. Integr. Equ. Oper. Theory **13/3**, 307–315 (1990)
3. W.O. Amrein, D.B. Pearson, Commutators of Hilbert-Schmidt type and the scattering cross section. Ann. Inst. H. Poincaré Sect. A (N.S.) **30/2**, 89–107 (1979)
4. H. Araki, S. Yamagami, An inequality for Hilbert-Schmidt norm. Commun. Math. Phys. **81/1**, 89–96 (1981)
5. A. Bachir, Fuglede-Putnam theorem for w-hyponormal or class $\mathcal{Y}$ operators. Ann. Funct. Anal. **4/1**, 53–60 (2013)
6. A. Bachir, F. Lombarkia, Fuglede-Putnam's theorem for W-hyponormal operators. Math. Inequal. Appl. **15/4**, 777–786 (2012)
7. A. Bachir, T. Prasad, Fuglede-Putnam theorem for (α, β)-normal operators. Rend. Circ. Mat. Palermo (2) **69/3**, 1243–1249 (2020)
8. A. Bachir, A. Segres, An asymmetric Fuglede-Putnam's theorem and orthogonality. Kyungpook Math. J. **46/4**, 497–502 (2006)
9. A.N. Bakir, S. Mecheri, Another version of Fuglede-Putnam theorem. Georgian Math. J. **16/3**, 427–433 (2009)
10. J. Barría, The commutative product $V_1^* V_2 = V_2 V_1^*$ for isometries V_1 and V_2. Indiana Univ. Math. J. **28**, 581–585 (1979)
11. S.K. Berberian, *Introduction to Hilbert space*. Reprinting of the 1961 original. With an addendum to the original (Chelsea Publishing Co., New York, 1976)
12. S.K. Berberian, Extensions of a theorem of Fuglede and Putnam. Proc. Am. Math. Soc. **71/1**, 113–114 (1978)
13. J.B. Conway, *A Course in Functional Analysis*, 2nd edn. (Springer, New York, 1990)
14. J.B. Conway, W. Szymański, Linear combinations of hyponormal operators. Rocky Mount. J. Math. **18/3**, 695–705 (1988)
15. S.S. Dragomir, M.S. Moslehian, Some inequalities for (α, β)-normal operators in Hilbert spaces. Facta Univ. Ser. Math. Inform. **23**, 39–47 (2008)
16. B.P. Duggal, On generalised Putnam-Fuglede theorems. Monatsh. Math. **107/4**, 309–332 (1989)
17. B.P. Duggal, A remark on generalised Putnam-Fuglede theorems. Proc. Am. Math. Soc. **129/1**, 83–87 (2001)
18. B.P. Duggal, C.S. Kubrusly, A Putnam-Fuglede commutativity property for Hilbert space operators. Linear Algebra Appl. **458**, 108–115 (2014)
19. Ju.B. Farforovskaja, Example of a Lipschitz function of self-adjoint operators that gives a nonnuclear increment under a nuclear perturbation (English). J. Sov. Math. **4**, 426–433 (1975–1976)
20. Ju.B. Farforovskaja, An estimate of the difference $f(B) - f(A)$ in the classes C_p (English). J. Sov. Math. **8**, 146–148 (1977)
21. Ju.B. Farforovskaya, An estimate of the norm $\| f(B) - f(A) \|$ for self-adjoint operators A and B (English). J. Sov. Math. **14**, 1133–1149 (1980)
22. T. Furuta, On relaxation of normality in the Fuglede-Putnam theorem. Proc. Am. Math. Soc. **77**, 324–328 (1979)
23. T. Furuta, An extension of the Fuglede-Putnam theorem to subnormal operators using a Hilbert-Schmidt norm inequality. Proc. Am. Math. Soc. **81/2**, 240–242 (1981)
24. T. Furuta, A Hilbert-Schmidt norm inequality associated with the Fuglede-Putnam theorem. Bull. Aust. Math. Soc. **25/2**, 177–185 (1982)
25. F. Gao, X. Fang, The Fuglede-Putnam theorem and Putnam's inequality for quasi-class (A, k) operators. Ann. Funct. Anal. **2/1**, 105–113 (2011)

26. W. Gong, A simple proof of an extension of the Fuglede-Putnam theorem. Proc. Am. Math. Soc. **100/3**, 599–600 (1987)

27. S. Jo, Y. Kim, E. Ko, On Fuglede-Putnam properties. Positivity **19/4**, 911–925 (2015)

28. D.R. Jocić, Integral representation formula for generalized normal derivations. Proc. Am. Math. Soc. **127/8**, 2303–2314 (1999)

29. H. Kamowitz, On operators whose spectrum lies on a circle or a line. Pacif. J. Math. **20**, 65–68 (1967)

30. F. Kittaneh, Commutators of C_p type. Thesis (Ph.D.)-Indiana University. *ProQuest LLC, Ann Arbor, MI*, 1982

31. F. Kittaneh, On generalized Fuglede-Putnam theorems of Hilbert-Schmidt type. Proc. Am. Math. Soc. **88/2**, 293–298 (1983)

32. F. Kittaneh, On Lipschitz functions of normal operators. Proc. Am. Math. Soc. **94/3**, 416–418 (1985)

33. V. Lauric, A note on subnormal and hyponormal derivations. Kyungpook Math. J. **48/2**, 281–286 (2008)

34. S. Mecheri, An extension of the Fuglede-Putnam theorem to (p, k)-quasihyponormal operators. Sci. Math. Jpn. **62/2**, 259–264 (2005)

35. S. Mecheri, K. Tanahashi, A. Uchiyama, Fuglede-Putnam theorem for p-hyponormal or class $\mathcal{Y}$ operators. Bull. Korean Math. Soc. **43/4**, 747–753 (2006)

36. S. Mecheri, A. Uchiyama, An extension of the Fuglede-Putnam's theorem to class A operators. Math. Inequal. Appl. **13/1**, 57–61 (2010)

37. R.L. Moore, D.D. Rogers, T.T. Trent, A note on intertwining M-hyponormal operators. Proc. Am. Math. Soc. **83/3**, 514–516 (1981)

38. R.L. Moore, G. Weiss, The metric Fuglede property and normality. Can. J. Math. **35/3**, 516–525 (1983)

39. M.H. Mortad, Yet more versions of the Fuglede-Putnam theorem. Glasgow Math. J. **51/3**, 473–480 (2009)

40. M.H. Mortad, *An Operator Theory Problem Book* (World Scientific Publishing Co., Singapore, 2018)

41. M.H. Mortad, On the absolute value of the product and the sum of linear operators. Rend. Circ. Mat. Palermo, II. Ser **68/2**, 247–257 (2019)

42. M.H. Mortad, *Counterexamples in Operator Theory* (Birkhäuser/Springer, Cham, 2022)

43. M.S. Moslehian, S.M.S. Nabavi Sales, Fuglede-Putnam type theorems via the Aluthge transform. Positivity **17/1**, 151–162 (2013)

44. T. Okuyama, K. Watanabe, The Fuglede-Putnam Theorem and a Generalization of Barría's Lemma. Proc. Am. Math. Soc. **126/9**, 2631–2634 (1998)

45. P. Pagacz, The Putnam-Fuglede property for paranormal and *-paranormal operators. Opuscula Math. **33/3**, 565–574 (2013)

46. S.M. Patel, A note on p-hyponormal operators for $0 < p < 1$. Integr. Equ. Oper. Theor. **21/4**, 498–503 (1995)

47. S.M. Patel, On intertwining p-hyponormal operators. Indian J. Math. **38/3**, 287–290 (1996)

48. M. Radjabalipour, On majorization and normality of operators. Proc. Am. Math. Soc. **62/1**, 105–110 (1977)

49. M. Radjabalipour, An extension of Putnam-Fuglede theorem for hyponormal operators. Math. Z. **194/1**, 117–120 (1987)

50. M. Reed, B. Simon, *Methods of Modern Mathematical Physics*, vol. 1. Functional Analysis (Academic Press, Cambridge, 1980)

51. R. Schatten, Norm ideals of completely continuous operators. Second printing. Ergebnisse der Mathematik und ihrer Grenzgebiete, Band 27 (Springer, Berlin, New York, 1970)

52. V. Shulman, Some remarks on the Fuglede-Weiss theorem. Bull. London Math. Soc. **28/4**, 385–392 (1996)

53. V. Shulman, L. Turowska, Operator synthesis. II. Individual synthesis and linear operator equations. J. Reine Angew. Math. **590**, 143–187 (2006)

54. J.G. Stampfli, Hyponormal operators. Pacif. J. Math. **12**, 1453–1458 (1962)

55. J.G. Stampfli, B.L. Wadhwa, On dominant operators. Monatsh. Math. **84/2**, 143–153 (1977)
56. K. Tanahashi, On log-hyponormal operators. Integr. Equ. Oper. Theor. **34/3**, 364–372 (1999)
57. K. Tanahashi, S.M. Patel, A. Uchiyama, On extensions of some Flugede-Putnam type theorems involving (p, k)-quasihyponormal, spectral, and dominant operators. [On extensions of some Fuglede-Putnam type theorems involving (p, k)-quasihyponormal, spectral, and dominant operators.] Math. Nachr. **282/7**, 1022–1032 (2009)
58. A. Uchiyama, Berger-Shaw's theorem for p-hyponormal operators. Integr. Equ. Oper. Theor. **33/2**, 221–230 (1999)
59. A. Uchiyama, K. Tanahashi, Fuglede-Putnam's theorem for p-hyponormal or log-hyponormal operators. Glasg. Math. J.. **44/3**, 397–410 (2002)
60. D. Voiculescu, Some results on norm-ideal perturbations of Hilbert space operators. J. Oper. Theory **2/1**, 3–37 (1979)
61. G. Weiss, The Fuglede commutativity theorem modulo the Hilbert-Schmidt class and generating functions for matrix operators. II. J. Oper. Theor. **5/1**, 3–16 (1981)
62. T. Yoshino, Remark on the generalized Putnam-Fuglede theorem. Proc. Am. Math. Soc. **95/4**, 571–572 (1985)

Chapter 3
Asymptotic Versions

Among possible generalizations of the Fuglede theorem, we have the so-called asymptotic (or approximate) versions. For example, if $X, N \in B(H)$ where N is normal, and $\|NX - XN\|$ is small, is $\|N^*X - XN^*\|$ small? Is $\|E(\Delta)X - XE(\Delta)\|$ small where $E(\Delta)$ is the spectral measure associated with N, and Δ is an arbitrary Borel set in the complex plane?

J.J. Bastian and K.J. Harrison [2] showed that the second question is untrue. More precisely, they came up with a normal N, a Borel set Δ and a sequence X_n with $\|X_n\| = 1$ for all n and $\|NX_n - X_nN\| \to 0$, yet

$$\|E(\Delta)X_n - X_nE(\Delta)\| = 1$$

for all n.

Then B.E. Johnson and J.P. Williams [5] found an example that refutes the first question. More specifically, they have shown the existence of a normal N and a sequence X_n such that $\|NX_n - X_nN\| \to 0$ but $\|N^*X_n - X_nN^*\| \geq 1$ for all n.

Apparently, the pioneering asymptotic version is due to R. Moore, who observed that in the aforementioned example, the operators (X_n) were not uniformly bounded. This observation has guided him to add a control on the norm of X. He then obtained:

Theorem 3.0.7 ([6]) *Let $N \in B(H)$ be normal. For each $\varepsilon > 0$, there is $\delta > 0$ such that $\|X\| \leq 1$ and $\|NX - XN\| < \delta$ imply that $\|N^*X - XN^*\| < \varepsilon$.*

Remark The condition $\|X\| \leq 1$ can be replaced by $\|X\| \leq \alpha$ for a certain α.

The proof, which is borrowed from [6], uses a somehow Rosenblum-like argument.

© The Author(s), under exclusive license to Springer Nature Switzerland AG 2022
M. H. Mortad, *The Fuglede-Putnam Theory*, Lecture Notes in Mathematics 2322,
https://doi.org/10.1007/978-3-031-17782-8_3

Proof WLOG, suppose $\|N\| \le 1$, and set $\|NX - XN\| = \eta$. If $k \in \mathbb{N}$, then it can be shown by induction that

$$\|N^k X - XN^k\| \le k\|N\|^{k-1}\|NX - XN\|$$

(for *any* $X, N \in B(H)$). Hence

$$\|N^k X - XN^k\| \le k\eta$$

still for any positive integer k. So, for any complex λ,

$$\left\|e^{i\bar{\lambda}N} X - Xe^{i\bar{\lambda}N}\right\| = \left\|\sum_{k=0}^{\infty} \frac{(i\bar{\lambda})^k}{k!}(N^k X - XN^k)\right\| \le \sum_{k=0}^{\infty} \frac{|\lambda|^k}{k!}\|N^k X - XN^k\|$$

$$\le \sum_{k=0}^{\infty} \frac{|\lambda|^k}{k!}k\eta = \eta|\lambda|\sum_{k=0}^{\infty}\frac{|\lambda|^{k-1}}{(k-1)!}$$

$$= \eta|\lambda|e^{|\lambda|}.$$

By the same token, it may be shown that $e^{|\lambda|}$ bounds each of $\|e^{i\bar{\lambda}N}\|$, $\|e^{-i\lambda N^*}\|$ and $\|e^{i\lambda N^*}\|$. Therefore

$$\left\|e^{i\lambda N^*}(e^{i\bar{\lambda}N} X - Xe^{i\bar{\lambda}N})e^{-i\bar{\lambda}N}e^{-i\lambda N^*}\right\| \le \eta|\lambda|e^{4|\lambda|}.$$

Since $NN^* = N^*N$, it ensues that

$$\left\|e^{i(\lambda N^* + \bar{\lambda}N)} Xe^{-i(\lambda N^* + \bar{\lambda}N)} - e^{i\lambda N^*} Xe^{-i\lambda N^*}\right\| \le \eta|\lambda|e^{4|\lambda|}$$

whereby

$$\left\|e^{i\lambda N^*} Xe^{-i\lambda N^*}\right\| \le \left\|e^{i(\lambda N^* + \bar{\lambda}N)} Xe^{-i(\lambda N^* + \bar{\lambda}N)}\right\| + \eta|\lambda|e^{4|\lambda|}.$$

By the self-adjointness of $\lambda N^* + \bar{\lambda}N$, we obtain the "unitarity" of $e^{\pm i(\lambda N^* + \bar{\lambda}N)}$. Accordingly,

$$\left\|e^{i\lambda N^*} Xe^{-i\lambda N^*}\right\| \le \|X\| + \eta|\lambda|e^{4|\lambda|} \le 1 + \eta|\lambda|e^{4|\lambda|}.$$

Now, define $\lambda \mapsto \varphi(\lambda) = e^{i\lambda N^*} Xe^{-i\lambda N^*}$, which is an analytic operator-valued function.

As is known (say in [1]), the Hahn-Banach theorem allows us to carry over many results in basic complex analysis to vector-valued versions. One of these results is Cauchy's integral formula. By the latter, we have

$$\varphi'(0) = \frac{1}{2\pi i} \int_{\Gamma} \frac{\varphi(\lambda)}{\lambda^2} d\lambda,$$

where Γ is some closed Jordan curve whose interior contains the origin. In particular, when Γ is a circle of radius r centered at the origin, we get

$$\|\varphi'(0)\| \leq \frac{1}{r} \max_{\Gamma} \|\varphi(\lambda)\| \leq \frac{1}{r}(1 + \eta r e^{4r}) = \frac{1}{r} + \eta e^{4r}.$$

Since $\varphi'(\lambda) = i N^* e^{i\lambda N^*} X e^{-i\lambda N^*} - i e^{i\lambda N^*} X N^* e^{-i\lambda N^*}$, it ensues that $\|\varphi'(0)\| = \|N^* X - X N^*\|$. Whence

$$\|N^* X - X N^*\| \leq \frac{1}{r} + \eta e^{4r}$$

for all $r > 0$. So if $\varepsilon > 0$, then choosing $r = 2/\varepsilon$ and $\delta = 1/2\varepsilon e^{-4r}$. Hence, if $\|NX - XN\| < \delta$, i.e., $\eta \leq \delta$, we obtain

$$\|N^* X - X N^*\| \leq \varepsilon,$$

as needed.   □

By Berberian's trick, the following Putnam-like generalization is easily obtained.

Theorem 3.0.8 ([6]) *Let $N, M \in B(H)$ be normal. For each $\varepsilon > 0$, there is $\delta > 0$ such that $\|X\| \leq 1$ and $\|NX - XM\| < \delta$ imply that $\|N^* X - X M^*\| < \varepsilon$.*

As another consequence of Theorem 3.0.7, we have the ensuing result:

Corollary 3.0.9 ([6]) *Let $N, M \in B(H)$ be normal and let ψ be a complex-valued continuous function defined on the union of the spectra of N and M. For each $\varepsilon > 0$, there is $\delta > 0$ such that $\|X\| \leq 1$ and $\|NX - XM\| < \delta$ imply that $\|\psi(N)X - X\psi(M)\| < \varepsilon$.*

Remark R. Moore's result may be formulated as follows: for each neighborhood E of 0 in the norm topology, there is a neighborhood D of 0 in the same topology such that the conditions $\|X\| \leq 1$ and $NX - XN \in D$ imply $N^* X - X N^* \in E$. D.D. Rogers then extended R. Moore's result to the strong and weak operator topologies. We state this result without proof.

Theorem 3.0.10 ([7]) *Let $N, M \in B(H)$ be normal and let ψ be a complex-valued continuous function defined on the union of the spectra of N and M. If E is a neighborhood of 0 in the strong operator topology (resp. weak operator topology), there is a neighborhood D of 0 in the strong topology (resp. weak topology) such*

that the conditions $\|X\| \leq 1$ *and* $NX - XM \in D$ *imply that* $\psi(N)X - X\psi(M) \in E$
(in particular, one has $N^*X - XM^* \in E$ *).*

T. Furuta generalized the previous results to the case of subnormality:

Theorem 3.0.11 ([3]) *Let* $A, B^* \in B(H)$ *be both subnormal. If* E *is a neighborhood of 0 in the norm topology (resp. strong operator topology, weak operator topology), there is a neighborhood* D *of 0 in the norm topology (resp. strong operator topology, weak operator topology) such that the conditions* $\|X\| \leq 1$ *and* $AX - XB \in D$ *imply* $A^*X - XB^* \in E$.

Remark The authors in [4] studied a similar problem on a more general structure than $B(H)$.

Remark T. Furuta asked in [3] whether his result may be generalized to hyponormality? However, no answer is available to me.

References

1. W. Arendt, C.J.K. Batty, M. Hieber, F. Neubrander, *Vector-Valued Laplace Transforms and Cauchy Problems*, vol. **96**, 2nd edn. Monographs in Mathematics (Birkhäuser/Springer Basel AG, Basel, 2011)
2. J.J. Bastian, K.J. Harrison, Subnormal weighted shifts and asymptotic properties of normal operators. Proc. Am. Math. Soc. **42**, 475–479 (1974)
3. T. Furuta, Normality can be relaxed in the asymptotic Fuglede-Putnam theorem. Proc. Am. Math. Soc. **79/4**, 593–596 (1980)
4. E.A. Gorin, M.I. Karahanjan, An asymptotic variant of the Fuglede-Putnam commutator theorem for the elements of a Banach algebra (Russian). Mat. Zametki **22/2**, 179–188 (1977)
5. B.E. Johnson, J.P. Williams, The range of a normal derivation. Pacif. J. Math. **58/1**, 105–122 (1975)
6. R.L. Moore, An asymptotic Fuglede theorem. Proc. Am. Math. Soc. **50**, 138–142 (1975)
7. D.D. Rogers, On Fuglede's theorem and operator topologies. Proc. Am. Math. Soc. **75/1**, 32–36 (1979)

Chapter 4
Generalizations of the Fuglede-Putnam Theorem to Banach Algebras and Spaces

4.1 Generalizations of the Fuglede-Putnam Theorem to Banach Algebras

First, we recall briefly the notions of Banach and C^*-algebras.

Definition 4.1.1 An algebra over $\mathbb{C}$, is a $\mathbb{C}$-vector space $\mathcal{A}$ such that if we define a multiplication $\cdot : \mathcal{A} \times \mathcal{A} \to \mathcal{A}$, then $(A, +, \cdot)$ becomes a ring, with a unit element noted e, such that also

$$\alpha(ab) = (\alpha a)b = a(\alpha b), \ \forall a, b \in \mathcal{A}, \alpha \in \mathbb{C}.$$

A normed algebra is an algebra $\mathcal{A}$ endowed with a norm $\| \cdot \|$ such that $\|ab\| \leq \|a\|\|b\|$ for all $a, b \in \mathcal{A}$, and $\|e\| = 1$.

If a normed algebra is complete, then we call it a Banach algebra.

Definition 4.1.2 A C^*-algebra is a complex Banach algebra $\mathcal{A}$, along with a map $a \mapsto a^*$ of $\mathcal{A}$ into $\mathcal{A}$ with the following properties:

$$(a + b)^* = a^* + b^*, \ (\alpha a)^* = \overline{\alpha} a^*, \ (ab)^* = b^* a^*, \ (a^*)^* = a, \text{ and } \|a^* a\| = \|a\|^2$$

for all $a, b \in \mathcal{A}$ and all $\alpha \in \mathbb{C}$.

Remark A map $a \mapsto a^*$ of $\mathcal{A}$ into $\mathcal{A}$ with the following properties:

$$(a + b)^* = a^* + b^*, \ (\alpha a)^* = \overline{\alpha} a^*, \ (ab)^* = b^* a^*, \text{ and } (a^*)^* = a$$

is called an involution.

For more about standard facts and results about Banach algebras et al., readers could consult [16] or [19].

© The Author(s), under exclusive license to Springer Nature Switzerland AG 2022
M. H. Mortad, *The Fuglede-Putnam Theory*, Lecture Notes in Mathematics 2322,
https://doi.org/10.1007/978-3-031-17782-8_4

Is there a Fuglede-Putnam theorem over C^*-algebras? The answer is yes. First, recall that an element $a \in \mathcal{A}$, where $\mathcal{A}$ is a C^*-algebra, is said to be self-adjoint when $a^* = a$. It is called normal if $aa^* = a^*a$, and unitary when $a^* = a^{-1}$. The same definitions remain valid over Banach algebras equipped with an involution.

Here is the generalization to C^*-algebras, whose proof is omitted for it is very much the same as Rosenblum's in $B(H)$.

Theorem 4.1.1 (See, e.g., [16] or [3]) *Let $\mathcal{A}$ be a C^*-algebra, and let $a, b, c \in \mathcal{A}$ be such that $ab = ca$, where b and c are normal elements. Then $ab^* = c^*a$.*

Remark Do we have a Fuglede-Putnam theorem on general Banach algebras with an involution? The answer, this time, is negative. W. Rudin supplied a counterexample in Exercise 28, Chap. 12 in [19]. There, he gave two *normal* elements x and y such that $xy = yx$, whereas $xy^* \neq y^*x$. Readers might be interested in reading a related paper: [2].

Remark More Fuglede-Putnam versions in certain particular rings may be consulted in [3].

4.2 Generalizations of the Fuglede-Putnam Theorem to Banach Spaces

First, readers should be aware that there is, alas, no spectral theorem in Banach spaces, and also that non-Hilbert Banach spaces are not complemented. Among other impediments to the extension of certain results on Hilbert spaces to Banach spaces is the fact that the notion of the adjoint on Banach spaces is not very practical (see, e.g., [18] for its definition for bounded operators and [13] for unbounded ones).

Thankfully, it is possible to define normal operators without using the notion of the adjoint, yet this definition fully tally with the usual definition over Hilbert spaces.

To obtain versions on Banach spaces, let us first reformulate the Fuglede-Putnam theorem in terms of Cartesian decompositions. Recall that any $T \in B(H)$ (where H is still a Hilbert space) is expressible as $T = R + iS$, where $R, S \in B(H)$ are self-adjoint. As is known, R is called the real part of T while S is called the imaginary part of T. It is straightforward to check that T is normal if and only if $RS = SR$.

Let $X \in B(H)$ be such that $XA = BX$, where $A, B \in B(H)$ are normal. If we write $A = C + iD$ and $B = E + iF$, where C, D, E, F are all self-adjoint, then $CD = DC$ and $EF = FE$ by the normality of A and B respectively. The Fuglede-Putnam theorem says that $XA = BX \Leftrightarrow XA^* = B^*X$. In other words,

$$X(C + iD) = (E + iF)X \Longleftrightarrow X(C - iD) = (E - iF)X,$$

which, in turn, is equivalent to $XC - EX = 0$ and $FX - XD = 0$.

Now, we ask: What does normality mean in a Banach space? What does self-adjointness mean? The following definitions extend notions already known in Hilbert spaces. Readers are referred to [5] for more properties. Also, see [11] and [12].

Definition 4.2.1 Let X be a complex Banach space and let $A \in B(X)$. Say that A is Hermitian if $\|e^{itA}\| = 1$ for all $t \in \mathbb{R}$. An operator $N \in B(X)$ is said to be normal if $N = A + iB$ where $A, B \in B(H)$ are two commuting Hermitian operators.
    We may then write symbolically $N^* = A - iB$.

Remark There are equivalent definitions to "hermitianity". We may mention Vidav's definition (which appeared in [20]). Another definition is based upon the numerical range. Recall first that the numerical range of $T \in B(X)$, where X is a Banach space, is defined by

$$V(T) = \{f(Tx) : x \in X, f \in X^*, \|x\| = \|f\| = f(x) = 1\},$$

where X^* is the dual space of X. Once this definition is known, it may be shown that A is Hermitian if and only if $V(A) \subset \mathbb{R}$. See p. 46 of [5] for proofs of the last claims.

Remark Readers familiar with self-adjoint and normal operators over Hilbert spaces should be wary of possible confusion with these new concepts. For instance, a normal operator N on a Banach space X does not have to satisfy $\|Nx\| = \|N^*x\|$ for all $x \in X$. A counterexample appeared in [8], along with a result about when this equality is true for a normal N.

To get to the generalization of the Fuglede-Putnam theorem, we introduce two other useful notions. Let $A \in B(X)$, where X is a Banach space, and define $L_A : B(X) \rightarrow B(X)$ by $L_A X = AX$ (this is called the left multiplication). Similarly, we may define the right multiplication: $R_A(X) = XA$. Clearly, $L_A, R_A \in B[B(X)]$. Moreover, it is seen (see, e.g., [7], C^*-algebra Chapter) that

$$\|L_A\| = \|R_A\| = \|A\|.$$

It is easy to see that when A and B are Hermitian, then so are L_A and R_B, as is $L_A - R_B$. More details may be found in [9].
    We have already observed that $XA = BX$, where $X \in B(H)$ and $A, B \in B(H)$ are normal, is equivalent to $XC - EX = 0$ and $FX - XD = 0$, where $A = C + iD$, $B = E + iF$, $CD = DC$ and $EF = FE$. So, define on $B(H)$ the linear operators $\mathcal{A}$ and $\mathcal{B}$ by

$$\mathcal{A}X = XC - EX \text{ and } \mathcal{B}X = FX - XD.$$

Therefore, the Fuglede-Putnam theorem may be formulated as follows: If $X \in B(H)$, then

$$(\mathcal{A} + i\mathcal{B})X = 0 \Longrightarrow \mathcal{A}X = \mathcal{B}X = 0.$$

In the above form, the Fuglede-Putnam theorem has an extension to arbitrary Banach spaces.

Theorem 4.2.1 *Let X be a complex Banach space. If $x \in X$, and $N = A + iB$ is normal (and so A and B are commuting Hermitian operators), then*

$$Nx = 0 \Longrightarrow Ax = Bx = 0.$$

Proof *([10])* By hypothesis, we have that $Bx = iAx$. Since $AB = BA$, we may show by induction that $B^n x = (iA)^n x$ for all $n \in \mathbb{N}$. Hence $e^{\lambda B}x = e^{i\lambda A}x$ for all complex λ and all x. Writing $\lambda = \alpha + i\beta$, where $\alpha, \beta \in \mathbb{R}$, we see that

$$e^{\lambda B}x = e^{i\beta B}e^{\alpha B}x = e^{i\beta B}e^{i\alpha A}x,$$

thereby $\|e^{\lambda B}x\| \le \|x\|$. Thus, $x \mapsto e^{\lambda B}x$ is a constant function by Liouville's theorem. By differentiating the previous function at $\lambda = 0$, we obtain $e^{\lambda B}Bx = 0$, and so $Bx = 0$ and also $Ax = 0$, as needed.   □

The following generalization of Theorem 4.2.1 was shown in [8]:

Theorem 4.2.2 *Let X be a complex Banach space. If $N = A + iB$ is normal (i.e., A and B are commuting Hermitian operators), and if for some $x \in X$, we have $N^2 x = 0$, then $Ax = Bx = 0$.*

Next, we show a Fuglede-Putnam theorem for a class of hyponormal operators on Banach spaces, but first, we define hyponormal operators on Banach spaces.

Definition 4.2.2 ([14]) Let X be a Banach space. Say that $T \in B(X)$ is hyponormal provided there are Hermitian operators A and B such that $T = A + iB$ and $i(AB - BA)$ is positive, i.e., its numerical range is a subset of $[0, \infty)$.

Remark Observe that because A and B are Hermitian, then $i(AB - BA)$ is already Hermitian, and so its numerical range is a subset of the real line (for a proof see, e.g., p. 47 in [5]). Hence the above definition makes sense.

Remark Again, those familiar with hyponormal operators on Hilbert spaces should be aware that if $T = A + iB \in B(X)$ is hyponormal, where X is a Banach space, then we *do not* necessarily have

$$\|(A - iB)x\| \le \|(A + iB)x\|$$

$(x \in X)$. A counterexample appeared in [8], where the writers gave an example of a normal operator (a fortiori hyponormal) that does not satisfy the previous inequality.

K. Mattila introduced in [15] the class of *-hyponormal operators. By a *-hyponormal operator, she meant a $T = A + iB \in B(X)$, where X is a Banach space, such that $\|e^{zT^*}e^{-\bar{z}T}\| \leq 1$ holds for all complex numbers z (where $T^* = A - iB$). Normal operators are *-hyponormal. K. Mattila showed in the above reference that a *-hyponormal operator is hyponormal, and the converse need not hold by supplying a counterexample.

We digress a little bit to recall that the simple fact that $e^{i\lambda N^*}e^{i\bar{\lambda}N}$, $\lambda \in \mathbb{C}$, are unitary, when N is normal, hence they have a norm equal to one, plays a crucial role in Rosenblum's proof of the Fuglede-Putnam theorem. The authors in [1] showed an analogous result in the case of subnormal operators, namely:

Lemma 4.2.3 *Let H be a Hilbert space, and let $A \in B(H)$ be subnormal. Then*

$$\left\| e^{i\lambda A^*}e^{i\bar{\lambda}A} \right\| \leq 1$$

for all $\lambda \in \mathbb{C}$.

We include the proof for readers' convenience.

Proof *([1])* As A is subnormal, there exist $B, C \in B(H)$ such that

$$N = \begin{pmatrix} A & B \\ 0 & C \end{pmatrix},$$

which is defined on $H \oplus H$, is normal. So, $e^{i\lambda N^*}e^{i\bar{\lambda}N}$, $\lambda \in \mathbb{C}$, are unitary, and for suitable $f(\lambda), g(\lambda) \in B(H)$, $\lambda \in \mathbb{C}$, one has

$$e^{i\lambda N} = \begin{pmatrix} e^{i\lambda A} & f(\lambda) \\ 0 & e^{i\lambda C} \end{pmatrix} \text{ and } e^{i\lambda N^*} = \begin{pmatrix} e^{i\lambda A^*} & 0 \\ g(\lambda) & e^{i\lambda C^*} \end{pmatrix}.$$

Hence

$$
\begin{aligned}
1 &= \sup_{\|x\|^2+\|y\|^2=1} \left\| e^{i(\lambda N^*+\bar{\lambda}N)} \begin{pmatrix} x \\ y \end{pmatrix} \right\| \\
&\geq \sup_{\|x\|=1} \left\| e^{i(\lambda N^*+\bar{\lambda}N)} \begin{pmatrix} x \\ 0 \end{pmatrix} \right\| \\
&= \sup_{\|x\|=1} \left(\|e^{i\lambda A^*}e^{i\bar{\lambda}A}x\|^2 + \|g(\lambda)e^{i\bar{\lambda}A}x\|^2 \right)^{1/2} \\
&\geq \left\| e^{i\lambda A^*}e^{i\bar{\lambda}A} \right\|,
\end{aligned}
$$

as wished. $\qquad\square$

Corollary 4.2.4 *Subnormal operators on a Hilbert space are* *-hyponormal.*

The generalization of the Fuglede-Putnam theorem to *-hyponormal operators on Banach spaces relies on the following lemma. Readers may consult [14] for a Fuglede-Putnam theorem for hyponormal operators on some particular Banach spaces.

Lemma 4.2.5 ([15]) *If* $T \in B(X)$ *is* *-hyponormal and $Tx = 0$ for some $x \in X$, then $T^*x = 0$.*

Theorem 4.2.6 ([15]) *Let X and Y be two Banach spaces. Let $T \in B(Y)$ be *-hyponormal and let $U \in B(X)$ be* *-hyponormal. If $TS = SU^*$ for a certain $S \in B(X, Y)$, then $T^*S = SU$.*

Proof *([15])* The main idea is to show that the operator defined by $S \mapsto \delta(S) = TS - SU^*$ is *-hyponormal on $B(X, Y)$. The result then follows by falling back on the foregoing lemma, and by noting that δ^* is the operator $S \mapsto T^*S - SU$.

Now, consider the left multiplication and the right multiplication operators $L_A S = AS$ and $R_B(S) = SB$ ($S \in B(X, Y)$), where $A \in B(Y)$ and $B \in B(X)$ are given. Observe that $L_A R_B = R_B L_A$ since for all S, one has

$$L_A R_B S = L_A(SB) = ASB = R_B(AS) = R_B L_A(S).$$

Hence $e^{L_A - R_B} = e^{L_A} e^{-R_B}$. Therefore, for each $S \in B(X, Y)$

$$e^{L_A - R_B}(S) = e^{L_A}[e^{-R_B}(S)] = e^{L_A}(Se^{-B}) = e^A Se^{-B}.$$

The previous relation combined with the assumption that T and U are *-hyponormal imply that for all $z \in \mathbb{C}$ and $S \in B(X, Y)$

$$\left\| e^{z\delta^*} e^{-\bar{z}\delta} S \right\| = \| e^{zT^*} e^{-\bar{z}T} Se^{\bar{z}U^*} e^{-zU} \| \leq \|S\|$$

which, upon passing to the supremum over the unit sphere in $B(X, Y)$, yields the desired result.   □

Remark For more related results about the Fuglede-Putnam theorem on complex Banach spaces, we refer readers to [4, 6, 8, 10, 17, 21].

References

1. S.T.M. Ackermans, S.J.L. van Eijndhoven, F.J. L. Martens, On almost commuting operators. Nederl. Akad. Wetensch. Indag. Math. **45**(4), 385–391 (1983)
2. M.V. Akhramovich, M.A. Muratov, V.S. Shul'man, Fuglede-Putnam theorem in algebras with involutions (Russian). Translated from Mat. Zametki **98**(4), 483–497 (2015), Math. Notes **98**(3–4), 537–549 (2015)

3. S.K. Berberian, Note on a theorem of Fuglede and Putnam. Proc. Am. Math. Soc. **10**, 175–182 (1959)

4. E. Berkson, H.R. Dowson, G.A. Elliott, On Fuglede's theorem and scalar-type operators. Bull. Lond. Math. Soc. **4**, 13–16 (1972)

5. F.F. Bonsall, J. Duncan, Numerical ranges of operators on normed spaces and of elements of normed algebras, in *London Mathematical Society Lecture Note Series*, vol. 2 (Cambridge University Press, London/New York, 1971)

6. K.N. Boyadzhiev, Commuting C_0 groups and the Fuglede-Putnam theorem. Stud. Math. **81**(3), 303–306 (1985)

7. J.B. Conway, *A Course in Functional Analysis*, 2nd edn. (Springer, Berlin, 1990)

8. M.J. Crabb, P.G. Spain, Commutators and normal operators. Glasg. Math. J. **18**(2), 197–198 (1977)

9. H.R. Dowson, T.A. Gillespie, P.G. Spain, A commutativity theorem for Hermitian operators. Math. Ann. **220**(3), 215–217 (1976)

10. C.K. Fong, Normal operators on Banach spaces. Glasg. Math. J. **20**(2), 163–168 (1979)

11. F.J. García-Pacheco, Selfadjoint operators on real or complex Banach spaces. Nonlinear Anal. **192**, 111696, 12 pp. (2020)

12. T.L. Gill, S. Basu, W.W. Zachary, V. Steadman, Adjoint for operators in Banach spaces. Proc. Am. Math. Soc. **132**(5), 1429–1434 (2004)

13. S. Goldberg, *Unbounded Linear Operators: Theory and Applications* (Dover Publications, Mineola, 2006)

14. K. Mattila, Complex strict and uniform convexity and hyponormal operators. Math. Proc. Camb. Philos. Soc. **96**(3), 483–493 (1984)

15. K. Mattila, A class of hyponormal operators and weak*-continuity of Hermitian operators. Ark. Mat. **25**(2), 265–274 (1987)

16. R. Meise, D. Vogt, in *Introduction to Functional Analysis*, Oxford Graduate Texts in Mathematics, vol. 2 (Oxford University Press, New York, 1997)

17. G. Mockenhaupt, W.J. Ricker, Fuglede's theorem, the bicommutant theorem and p-multiplier operators for the circle. J. Oper. Theory **49**(2), 295–310 (2003)

18. M. Reed, B. Simon, Methods of modern mathematical physics, in *Functional Analysis*, vol. 1 (Academic Press, New York, 1980)

19. W. Rudin, *Functional Analysis*, 2nd edn. (McGraw-Hill, New York, 1991)

20. I. Vidav, Eine metrische Kennzeichnung der selbstadjungierten Operatoren (German). Math. Z. **66**, 121–128 (1956)

21. B. Yood, Commutativity properties in Banach *-algebras. Pac. J. Math. **53**, 307–317 (1974)

Chapter 5
Generalizations to Unbounded Operators

5.1 All-Unbounded-Operator Versions

First, we tell readers that we are back here to Hilbert spaces.

The first versions of B. Fuglede and C.R. Putnam both dealt with unbounded operators. Nonetheless, not all operators involved were unbounded. Since the papers of B. Fuglede and C.R. Putnam, the first version apparently exclusively devoted to unbounded operators, is:

Theorem 5.1.1 ([16]) *Let N and T be strongly commuting normal operators and let A be a densely defined operator. Suppose that*

1. $D_0(T) \subset D(N) \subset D(A) \cap D(A^*)$[1] *and*
2. $\langle Nx, A^*y \rangle = \langle Ax, N^*y \rangle$ *for all* $x, y \in D_0(T)$.

Then the spectral measure of N, noted E, commutes with $\overline{A}$ *(the closure of A). That is,* $E(\Delta)\overline{A} \subset \overline{A}E(\Delta)$ *for all Borel sets* Δ.

The following all-unbounded version came up in 2003:

Theorem 5.1.2 ([10]) *If A is an unbounded self-adjoint operator and if N is an unbounded normal operator, then*

$$AN \subset N^*A \Longrightarrow AN^* \subset NA$$

whenever $D(N) \subset D(A)$.

A different proof of the previous theorem appeared later in [19].

---

[1] $D_0(T)$ is the union $\cup \operatorname{ran}[F(\Delta)]$ as Δ ranges over all bounded Borel sets, and F is the spectral measure associated with T.

© The Author(s), under exclusive license to Springer Nature Switzerland AG 2022
M. H. Mortad, *The Fuglede-Putnam Theory*, Lecture Notes in Mathematics 2322,
https://doi.org/10.1007/978-3-031-17782-8_5

Nevertheless, by scrutinizing the proof in [10], it is seen that only the closedness and the symmetricity of A were needed (self-adjointness was imposed for some subsequent results). The revised version of Theorem 5.1.2 should therefore be:

Theorem 5.1.3 ([10]) *If A is a closed symmetric operator and if N is an unbounded normal operator, then*

$$AN \subset N^*A \Longrightarrow AN^* \subset NA$$

whenever $D(N) \subset D(A)$.

Proof *([11])* Let P_{B_R} be the spectral projection associated with N. For convenience we set $H_R = \operatorname{ran} P_{B_R}$. Let us restrict A to the Hilbert space H_R, noted A_{H_R}. We claim that $A : H_R \to H_R$ and that A is bounded.

Clearly, H_R is a subset of $D(A)$ since $H_R \subset D(N)$ by the spectral theorem and $D(N) \subset D(A)$. On the other hand, since A_{H_R} is symmetric and everywhere defined, by the Hellinger-Toeplitz theorem (see, e.g., [14]), it is bounded on H_R.

Now, we show that $Ax \in H_R$ whenever $x \in H_R$. Let $x \in H_R$. By the spectral theorem, it is seen that

$$Ax \in H_R \iff \|(N^*)^k Ax\| \le \alpha R^k.$$

We also have $\|N^k x\| \le cR^k$ and since A is bounded, $\|AN^k x\| \le \alpha R^k$. However, for such an x we have

$$\|AN^k x\| = \|(N^*)^k Ax\|$$

as a consequence of the assumption, whereby $Ax \in H_R$.

Next, we show that $AN^* \subset NA$. To this end, let $x \in D(AN^*)$, then define $x_n = P_{B_n} x$. Since $P_{B_n} \to I$ in the strong operator topology, $x_n \to x$. Besides, $x_n \in D(AN^*)$ for both A and N^* are bounded operators on H_n. Let us now show that $AN^* x_n \to AN^* x$. Since A is symmetric and maps H_R into itself, H_R reduces A and hence

$$P_{B_R} A \subset A P_{B_R}.$$

It also reduces N by the spectral theorem, thereby

$$AN^* x_n = AN^* P_{B_n} x = P_{B_n} AN^* x \to AN^* x. \tag{5.1}$$

Now, we show that $x \in D(NA)$. Since both A and N are bounded on H_n, the Fuglede-Putnam for bounded operators gives

$$ANx_n = N^* Ax_n \Longrightarrow AN^* x_n = NAx_n.$$

This, combined with Eq. 5.1 yield $NAx_n \to AN^* x$.

Now, N maps $H_R^\perp = \operatorname{ran} P_{B_R^c}$ to $H_R^\perp$ (H_R is a reducing space for N) and N^{-1} is bounded on $H_R^\perp$ since in this case $N^{-1} = \int_{B_R^c} \frac{1}{\lambda} dP_\lambda$.

Moreover,

$$NAx_n - AN^*x = AN^*x_n - AN^*x \in H_R^\perp \text{ for } n > R,$$

so that if we apply the inverse of N, we get $Ax_n \to N^{-1}AN^*x$. The closedness of A gives $x \in D(A)$ and $Ax_n \to Ax$. But N too is closed and the convergence of $(NAx_n)_n$ together with $Ax_n \to Ax$ imply that

$$Ax \in D(N) \text{ (i.e, } x \in D(NA)) \text{ and } AN^*x = NAx,$$

as needed. $\qquad\qquad\qquad\qquad\qquad\qquad\qquad\qquad\qquad\qquad\qquad\quad$ □

By reading the previous proof thoroughly, we see that the following version may also be obtained by mimicking the same proof.

Theorem 5.1.4 *If A is a closed symmetric operator and if N is an unbounded normal operator, then*

$$AN \subset NA \Longrightarrow AN^* \subset N^*A,$$

whenever $D(N) \subset D(A)$.

Later in [13], the following generalization was obtained:

Theorem 5.1.5 *Let A be a closed operator with domain $D(A)$. Let M and N be two unbounded normal operators with domains $D(N)$ and $D(M)$ respectively. If $D(N) \subset D(AN) \subset D(A)$, then*

$$AN \subset MA \Longrightarrow AN^* \subset M^*A.$$

Remark The preceding theorem answers an open question from [12].

A very interesting generalization was obtained by F.C. Paliogiannis, under the assumptions $D(N) \subset D(A)$ and $D(M) \subset D(A^*)$. More precisely:

Theorem 5.1.6 ([20]) *Let A be a closed operator with domain $D(A)$. Let M and N be two unbounded normal operators with domains $D(N)$ and $D(M)$ respectively. If $D(N) \subset D(A)$ and $D(M) \subset D(A^*)$, then*

$$AN \subset MA \Longrightarrow AN^* \subset M^*A.$$

Remark Observe that Theorem 5.1.6 generalizes both Theorem 5.1.3 and Theorem 5.1.4, because when A is symmetric (and densely defined), then $D(A) \subset D(A^*)$.

Corollary 5.1.7 ([20]) *Let A be a closed operator with domain $D(A)$. Let N be an unbounded normal operator with domain with $D(N)$. If $D(N) \subset D(A) \cap D(A^*)$, then*

$$AN \subset NA \implies AN^* \subset N^*A.$$

Corollary 5.1.8 ([20]) *Let M and N be unbounded normal operators with domains $D(M)$ and $D(N)$ respectively. If $D(N) \subset D(M)$, then*

$$MN \subset NM \iff MN^* \subset N^*M.$$

Corollary 5.1.9 ([20]) *Let A, M and N be unbounded normal operators with domains $D(A)$, $D(M)$ and $D(N)$ respectively. If $D(N), D(M) \subset D(A)$, then*

$$AN \subset MA \iff AN^* \subset M^*A.$$

The ensuing result involves boundedly invertible normal operators.

Theorem 5.1.10 *[[2]] Let A, N and M be three boundedly invertible operators on a Hilbert space such that N and M are normal. If $AN = MA$, then*

$$A^*M = NA^* \text{ and } AN^* = M^*A.$$

Proof *([2])* We have

$$AN = MA \implies A^{-1}M \subset NA^{-1}.$$

Since A^{-1} is everywhere defined and bounded, the Fuglede-Putnam theorem gives $A^{-1}M^* \subset N^*A^{-1}$, and hence

$$M^*A \subset AN^*.$$

Applying adjoints on both sides implies that

$$NA^* \subset A^*M.$$

Now, as $A^{-1}M \subset NA^{-1}$, we get $N^{-1}A^{-1} \subset A^{-1}M^{-1}$. But all operators in the previous "inclusion" are everywhere defined and bounded. Therefore, we get

$$N^{-1}A^{-1} = A^{-1}M^{-1}.$$

Applying the Fuglede-Putnam theorem (the bounded version) yields

$$(N^{-1})^*A^{-1} = A^{-1}(M^{-1})^*.$$

Whence

$$(M^{-1})^*A \subset A(N^{-1})^*, \text{ and so } AN^* \subset M^*A.$$

Consequently, $A^*M \subset NA^*$. Thus $AN^* = M^*A$ and $A^*M = NA^*$, establishing the result. □

Remark It seems noteworthy to tell readers that a condition like $AN = NA$ may be obtained from $AN \subset NA$ if $A \in B(H)$ is invertible, and N is normal. Indeed, since $AN \subset NA$, it follows that $AE(\Delta) = E(\Delta)A$ for any Borel set Δ, where E is the spectral measure associated with N. Then $A^{-1}E(\Delta) = E(\Delta)A^{-1}$, which then implies that $A^{-1}N \subset NA^{-1}$. Thus, $NA \subset AN$, thereby $AN = NA$ (hence $A^{-1}N = NA^{-1}$, too), as needed. This argument is borrowed from J. Stochel and F.H. Szafraniec (Unbounded operators and subnormality, unpublished yet).

The coming result treats an interesting generalization that will allow us to get more versions involving unbounded operators only.

Theorem 5.1.11 ([6]) *If B is a bounded normal operator and if A is a densely defined operator with $\sigma(A) \neq \mathbb{C}$,[2] then:*

$$BA \subset AB \Longrightarrow g(B)A \subset Ag(B)$$

for any bounded complex Borel function g on $\sigma(B)$. In particular, we have $B^*A \subset AB^*$.

Proof *([6])* If $\lambda \in \rho(A)$, then

$$BA \subset AB \Longrightarrow B(A - \lambda I) \subset (A - \lambda I)B \Longrightarrow (A - \lambda I)^{-1}B \subset B(A - \lambda I)^{-1}.$$

Now, apply the Fuglede theorem (given that $(A - \lambda I)^{-1} \in B(H)$) to get

$$(A - \lambda I)^{-1}g(B) = g(B)(A - \lambda I)^{-1},$$

for any bounded complex Borel function g on $\sigma(B)$. Thus, $g(B)A \subset Ag(B)$. By choosing $g(z) = \bar{z}$, we obtain the particular claim of the theorem. □

The remaining portion of this section is borrowed from [4]. Below we give a generalization of Theorem 5.1.11 to a closable N, where only the normality of $\overline{N}$ is needed.

---

[2] Hence A is closed according to the definition adopted in this monograph.

Theorem 5.1.12 *Let p be a one variable complex polynomial. If N is a densely defined closable operator such that $\overline{N}$ is normal and if A is a densely defined operator with $\sigma[p(A)] \neq \mathbb{C}$, then*

$$NA \subset AN \implies N^*A \subset AN^*$$

whenever $D(A) \subset D(N)$.

Remark This is indeed a generalization of the bounded version of the Fuglede theorem. Observe that when $A, N \in B(H)$, then $\overline{N} = N$, $D(A) = D(N) = H$, and $\sigma[p(A)]$ is a compact set.

Proof First, we claim that $\sigma(A) \neq \mathbb{C}$, whereby A is closed. Let λ be in $\mathbb{C}\backslash\sigma[p(A)]$. Then, and as in [3], we obtain

$$p(A) - \lambda I = (A - \mu_1 I)(A - \mu_2 I) \cdots (A - \mu_n I)$$

for some complex numbers $\mu_1, \mu_2, \cdots, \mu_n$. By consulting again [3], readers see that $\sigma(A) \neq \mathbb{C}$.

Now, let $\lambda \in \rho(A)$. Then

$$NA \subset AN \implies NA - \lambda N \subset AN - \lambda N = (A - \lambda I)N.$$

Since $D(A) \subset D(N)$, it is seen that $NA - \lambda N = N(A - \lambda I)$. So

$$N(A - \lambda I) \subset (A - \lambda I)N \implies (A - \lambda I)^{-1}N \subset N(A - \lambda I)^{-1}.$$

Since $\overline{N}$ is normal, we may now apply Lemma 1.2.5 to get

$$(A - \lambda I)^{-1}N^* \subset N^*(A - \lambda I)^{-1}$$

because $(A - \lambda I)^{-1} \in B(H)$. Hence

$$N^*A - \lambda N^* \subset N^*(A - \lambda I) \subset (A - \lambda I)N^* = AN^* - \lambda N^*.$$

But

$$D(AN^*) \subset D(N^*) \text{ and } D(N^*A) \subset D(A) \subset D(N) \subset D(\overline{N}) = D(N^*).$$

Thus, $D(N^*A) \subset D(AN^*)$, and so

$$N^*A \subset AN^*,$$

as needed. $\square$

Now, we present a couple of consequences of the preceding result. The first one is given without proof.

Corollary 5.1.13 *If N is a densely defined closable operator such that $\overline{N}$ is normal and if A is an unbounded self-adjoint operator with $D(A) \subset D(N)$, then*

$$NA \subset AN \Longrightarrow N^*A \subset AN^*.$$

Corollary 5.1.14 *If N is a densely defined closable operator such that $\overline{N}$ is normal and if A is a boundedly invertible operator, then*

$$NA \subset AN \Longrightarrow N^*A \subset AN^*.$$

Proof We may write

$$NA \subset AN \Longrightarrow NAA^{-1} \subset ANA^{-1} \Longrightarrow A^{-1}N \subset NA^{-1}.$$

Since $A^{-1} \in B(H)$ and $\overline{N}$ is normal,

$$A^{-1}N^* \subset N^*A^{-1} \text{ and so } N^*A \subset AN^*,$$

as needed.

$\square$

A Fuglede-Putnam's version is impossible to obtain unless perhaps strong conditions are added. However, the following particular case of a possible Fuglede-Putnam's version is worth stating and proving. Besides, it is somewhat linked to the important notion of anti-commutativity (cf. [23]).

Proposition 5.1.15 *If N is a densely defined closable operator such that $\overline{N}$ is normal and if A is a densely defined operator with $\sigma(A) \neq \mathbb{C}$, then*

$$NA \subset -AN \Longrightarrow N^*A \subset -AN^*$$

whenever $D(A) \subset D(N)$.

Proof Consider

$$\widetilde{N} = \begin{pmatrix} N & 0 \\ 0 & -N \end{pmatrix} \text{ and } \widetilde{A} = \begin{pmatrix} 0 & A \\ A & 0 \end{pmatrix}$$

where $D(\widetilde{N}) = D(N) \oplus D(N)$ and $D(\widetilde{A}) = D(A) \oplus D(A)$. Then $\overline{\widetilde{N}}$ is normal and $\widetilde{A}$ is closed. Besides $\sigma(\widetilde{A}) \neq \mathbb{C}$. Now

$$\widetilde{N}\widetilde{A} = \begin{pmatrix} 0 & NA \\ -NA & 0 \end{pmatrix} \subset \begin{pmatrix} 0 & -AN \\ AN & 0 \end{pmatrix} = \widetilde{A}\widetilde{N}$$

for $NA \subset -AN$. Since $D(\widetilde{A}) \subset D(\widetilde{N})$, Theorem 5.1.12 applies, i.e., it gives $\widetilde{N}^*\widetilde{A} \subset \widetilde{A}\widetilde{N}^*$ which, upon examining their entries, yields the required result.   □

5.2   Other Generalizations to Unbounded Operators

It was conjectured in [9] that if T is an operator (densely defined and closed if necessary) and $B \in B(H)$ is normal, then

$$BT \subset TB^* \Longrightarrow B^*T \subset TB.$$

What made this conjecture interesting is the fact that it holds when $T \in B(H)$ for "$BT \subset TB^* \Rightarrow B^*T \subset TB$" then becomes "$BT = TB^* \Rightarrow B^*T = TB^*$", that is, we get back a bounded version of the Fuglede-Putnam theorem. As will be seen in the counterexamples section, this conjecture, as formulated, turned out to be false. Nonetheless, under some additional hypothesis, it was true.

Theorem 5.2.1 ([1]) *Let B be a bounded normal operator with a finite spectrum and let A be an unbounded operator on a complex Hilbert space H. Let $f, g : \mathbb{C} \to \mathbb{C}$ be two bounded Borel functions. Then*

$$BA \subset Af(B) \Longrightarrow g(B)A \subset A(g \circ f)(B).$$

Proof ([1]) Using induction we verify that $B^m A \subset A(f(B))^m$ for all $m \geq 0$. Therefore,

$$p(B)A \subset A(p \circ f)(B)$$

for any polynomial $p \in \mathbb{C}[X]$.

By assumption, B has a point spectrum with finitely many distinct eigenvalues $\lambda_j, j \in \{1, \cdots, n\}$, and corresponding eigenprojectors E_j adding up to the identity operator I, so $B = \sum_{j=1}^n \lambda_j E_j$ is the spectral representation of B. Now, using the Lagrange interpolation theorem, we find a polynomial p such that $p(\lambda_j) = g(\lambda_j)$ and $p(f(\lambda_j)) = g(f(\lambda_j))$ for all j. From the hypothesis $BA \subset Af(B)$, we obtain

$$g(B)A = p(B)A \subset A(p \circ f)(B) = A(g \circ f)(B).$$

□

Corollary 5.2.2 *With A and B as in Theorem 5.2.1, we have*

$$BA \subset AB^* \Longrightarrow B^*A \subset AB.$$

Proof Just apply Theorem 5.2.1 to the functions $f, g : z \mapsto \bar{z}$.   □

A similar reasoning applies to establish the following consequence. This also partially answers a question asked by P. E. T. Jørgensen in [8]. In the latter reference, some interesting observations may also be found.

Corollary 5.2.3 *With A and B as in Theorem 5.2.1, we likewise have*

$$BA \subset AB \Longrightarrow B^*A \subset AB^*.$$

Using Berberian's trick, we may generalize this result to the case of two normal operators whereby obtaining a Fuglede-Putnam version.

Proposition 5.2.4 ([1]) *Let B and C be bounded normal operators with a finite pure point spectrum and let A be an unbounded operator on a complex Hilbert space H. Then*

$$BA \subset AC \Longrightarrow B^*A \subset AC^*.$$

Proof *([1])* Define $\tilde{B}$ on $H \oplus H$ by:

$$\tilde{B} = \begin{pmatrix} B & 0 \\ 0 & C \end{pmatrix}$$

and let $\tilde{A} = \begin{pmatrix} 0 & A \\ 0 & 0 \end{pmatrix}$ with $D(\tilde{A}) = H \oplus D(A)$. Since $BA \subset AC$, it follows that

$$\tilde{B}\tilde{A} = \begin{pmatrix} 0 & BA \\ 0 & 0 \end{pmatrix} \subset \begin{pmatrix} 0 & AC \\ 0 & 0 \end{pmatrix} = \tilde{A}\tilde{B}$$

for $D(\tilde{B}\tilde{A}) = H \oplus D(A) \subset H \oplus D(AC) = D(\tilde{A}\tilde{B})$.

Now, since B and C are normal, so is $\tilde{B}$. Finally, apply Corollary 5.2.3 to the pair $(\tilde{B}, \tilde{A})$ to get

$$\tilde{B}^*\tilde{A} \subset \tilde{A}\tilde{B}^*$$

which, upon examining their entries, yields the required result.   □

5.3 Stochel's Version Involving Unbounded Subnormal and Hyponormal Operators

To the best of my knowledge, there are only a couple of papers or so dealing with the Fuglede-Putnam theorem in the case of unbounded closed hyponormal or subnormal operators, namely: [5] and [22] (we may add a somehow relevant paper which is [7]).

Apparently, T. Furuta was the first author to say a word on a Fuglede-Putnam theorem involving unbounded subnormal operators. He observed and stated *without proof* that the arguments used in his proof in [5] may be extended to the case of subnormal non-necessarily bounded operators so that one should obtain:

Theorem 5.3.1 *Let $X \in B(H)$. If A and B^* are unbounded subnormal operators, then*

$$XB \subset AX \Longrightarrow XB^* \subset A^*X.$$

Nevertheless, it seems that the only serious work in this direction is due to J. Stochel in [22]. So, the rest of this section is devoted to Stochel's results.

Getting to the main Fuglede-Putnam version requires several quite technical steps. We list some of these results without proofs that may be consulted in [22]. Readers are very much encouraged to look at Stochel's paper for further details.

The first result is a generalization of one that was established in [21] in a bounded setting.

Lemma 5.3.2 ([22]) *Let $X \in B(H, K)$ and let N be a normal operator in H, and let T be a closed hyponormal in K such that $XN \subset TX$. Then*

1. $|X|N \subset N|X|$,
2. $N = T$ if $K = H$, $X \geq 0$ and $\ker X = \{0\}$.

Before giving the next auxiliary result, we set up some notations. Let $X \in B(H, K)$. Write $\mathcal{I}(X) = \overline{\mathrm{ran}(X^*)} = \overline{\mathrm{ran}(|X|)}$ and $\mathcal{J}(X) = \overline{\mathrm{ran}(X)}$. The restriction of A to $\mathcal{I}(X)$ is denoted by A_X whereas the restriction of B to $\mathcal{J}(X)$ is designated by B^X. Let $X = i_X|X|$ be the polar decomposition of X in terms of partial isometries. Denote by $\widehat{X}$ the restriction of X to $\mathcal{I}(X)$. In the end, observe that it may be shown that $\widehat{X^*} = \widehat{X}^*$.

We shall also have need for the next two results:

Lemma 5.3.3 ([22]) *Let A and B be closed densely defined operators in H and K respectively, and let $X \in B(H, K)$ be such that $XA^* \subset BX$.*

1. *If $\mathcal{I}(X)$ reduces A, then B^X is closed densely defined in $\mathcal{J}(X)$ and*

$$\widehat{X}A_X^* \subset B^X\widehat{X}.$$

2. *If $\mathcal{I}(X)$ and $\mathcal{J}(X)$ reduce A and B to normal operators respectively, then*

$$XA \subset B^*X, \quad |X|A \subset A|X| \text{ and } |X^*|B \subset B|X^*|.$$

Proposition 5.3.4 ([22]) *Let $X \in B(H, K)$ and let N be a normal operator in H and let T be a closed hyponormal in K such that $XN \subset TX$. Then $\mathcal{I}(X)$ and $\mathcal{J}(X)$ reduce N and T to normal operators respectively.*

Here is the promised generalization.

Theorem 5.3.5 ([22]) *Let A be a closed subnormal (resp. a closed hyponormal) operator in H, and let B be a closed hyponormal (resp. a closed subnormal) operator in K. Let $X \in B(H, K)$ be such that $XA^* \subset BX$. Then*

1. $XA \subset B^*X$.
2. $|X|A \subset A|X|$.
3. $|X^*|B \subset B|X^*|$.
4. $\mathcal{I}(X)$ *reduces A to the normal* A_X.
5. $\mathcal{J}(X)$ *reduces B to the normal* B^X.

Proof ([22]) Assume in the first place that A is closed and subnormal whilst B is closed and hyponormal. Let N be a normal extension of A defined in a Hilbert space $H' \supset H$, then define $Y \in B(K, H')$ by $Yf = X^*f$ for $f \in K$. Let us show that $\mathcal{J}(X) = \mathcal{J}(Y^*)$. If $J \in B(H, H')$ is defined by $Jh = h$ for $h \in H$, then $Y = JX^*$. Hence $Y^* = XP$, where $P \in B(H', H)$ is the orthogonal projection of H' onto H. From the latter equality, we derive $\mathcal{J}(X) = \mathcal{J}(Y^*)$, as needed.

To show that (5) holds true, take adjoints in $XA^* \subset BX$ to obtain $X^*B^* \subset AX^*$, whereby $YB^* \subset NY$. Hence $Y^*N^* \subset BY^*$. By $\mathcal{J}(X) = \mathcal{J}(Y^*)$ and Proposition 5.3.4 we deduce that $\mathcal{J}(X) = \mathcal{J}(Y^*)$ reduces B to the normal operator B^X.

To show that (4) is true, consider $X^*B^* \subset AX^*$ to which we apply Lemma 5.3.3. We then obtain that $A_X = A^{X^*}$ is a closed densely defined operator in $\mathcal{I}(X)$. Besides,

$$\widehat{X}^*(B^X)^* \subset A_X\widehat{X}^*.$$

But A_X is subnormal (and a fortiori hyponormal) for $A_X \subset A \subset N$. By applying Proposition 5.3.4 to $\widehat{X}^*(B^X)^* \subset A_X\widehat{X}^*$, we see that $\mathcal{I}(X) = \mathcal{J}(\widehat{X}^*)$ reduces A_X to a normal operator, and so A_X is itself normal. So, $\mathcal{I}(X)$ reduces A to the normal operator A_X (see Proposition 1.3 in [22]). To obtain the other claims of the theorem, just apply Lemma 5.3.3.

Assume now that A is closed and hyponormal while B is closed and subnormal. Clearly, $XA^* \subset BX$ yields $X^*B^* \subset AX^*$. By the first part of the proof, $\mathcal{I}(X) = \mathcal{J}(X^*)$ reduces A to the normal A_X, and $\mathcal{J}(X) = \mathcal{I}(X^*)$ reduces B to the normal operator B^X. As before, and in order to get the rest of the conclusion, apply Lemma 5.3.3, and the proof is complete. □

The following consequence is interesting, and it generalizes results obtained in say [7, 17] and [18].

Corollary 5.3.6 ([22]) *Let A be a closed subnormal (resp. a closed hyponormal) operator in H, and let B be a closed hyponormal (resp. a closed subnormal) operator in K. Let $X \in B(H, K)$ be such that $XA^* \subset BX$. Then*

1. *A is normal provided* $\ker(X) = \{0\}$.
2. *B is normal provided* $\ker(X^*) = \{0\}$.

3. A^* and B are normal and unitarily equivalent if $\ker(X) = \{0\}$ and $\ker(X^*) = \{0\}$.
4. $B = A^*$ and A is normal when $K = H$, $X \geq 0$ and $\ker X = \{0\}$.

5.4 Counterexamples

First, we provide a boundedly invertible and positive self-adjoint unbounded operator A, and an unbounded normal operator N such that $AN^* = NA$ and $AN \not\subset N^*A$. In other terms, NA is self-adjoint while N^*A is not.

Example 5.4.1 ([13]) Define the following operators A and N by

$$Af(x) = (1 + |x|)f(x) \text{ and } Nf(x) = -i(1 + |x|)f'(x)$$

(with $i^2 = -1$) respectively on the domains

$$D(A) = \{f \in L^2(\mathbb{R}) : (1 + |x|)f \in L^2(\mathbb{R})\}$$

and

$$D(N) = \{f \in L^2(\mathbb{R}) : (1 + |x|)f' \in L^2(\mathbb{R})\}.$$

Clearly, A is self-adjoint on $D(A)$.

It can also be shown that N is normal on $D(N)$. We may find that

$$N^* f(x) = -i \, \text{sign}(x) f(x) - i(1 + |x|)f'(x)$$

with

$$D(N^*) = \{f \in L^2(\mathbb{R}) : (1 + |x|)f' \in L^2(\mathbb{R})\}$$

and where "sign" is the usual sign function.

Doing some arithmetic allows us to find that

$$AN^* f(x) = NAf(x) = -i(1 + |x|)\text{sign}(x)f(x) - i(1 + |x|)^2 f'(x)$$

for any f in the *equal* domains

$$D(AN^*) = D(NA) = \{f \in L^2(\mathbb{R}) : (1 + |x|)f \in L^2(\mathbb{R}), \ (1 + |x|)^2 f' \in L^2(\mathbb{R})\}$$

and so

$$NA = AN^* = (NA)^*$$

(since $A^{-1} \in B[L^2(\mathbb{R})]$). Nonetheless,

$$AN \not\subset N^*A$$

for

$$ANf(x) = -i(1 + |x|)^2 f'(x)$$

whereas

$$N^*Af(x) = -2i \operatorname{sign}(x)(1 + |x|)f(x) - i(1 + |x|)^2 f'(x),$$

that is

$$ANf(x) \neq N^*Af(x)$$

and regardless of the comparison of domains $D(AN)$ and $D(N^*A)$, this is sufficient to declare that $AN \not\subset N^*A$.

The previous example provides a counterexample to certain generalizations of the Fuglede-Putnam theorem. Next, we use it to refute a generalization of the Fuglede theorem via Berberian's trick. More precisely, we find a closed T and an unbounded normal M such that $TM \subset MT$ but neither $TM^* \subset M^*T$ nor $M^*T \subset TM^*$.

Example 5.4.2 ([15]) Consider

$$M = \begin{pmatrix} N^* & 0 \\ 0 & N \end{pmatrix} \text{ and } T = \begin{pmatrix} 0 & 0 \\ A & 0 \end{pmatrix}$$

where N is normal with domain $D(N)$ and A is closed with domain $D(A)$ and such that $AN^* = NA$ but $AN \not\subset N^*A$ and $N^*A \not\subset AN$. Clearly, M is normal and T is closed. Observe that $D(M) = D(N^*) \oplus D(N)$ and $D(T) = D(A) \oplus L^2(\mathbb{R})$. Now,

$$TM = \begin{pmatrix} 0 & 0 \\ A & 0 \end{pmatrix} \begin{pmatrix} N^* & 0 \\ 0 & N \end{pmatrix} = \begin{pmatrix} 0_{D(N^*)} & 0_{D(N)} \\ AN^* & 0 \end{pmatrix} = \begin{pmatrix} 0 & 0_{D(N)} \\ AN^* & 0 \end{pmatrix}$$

where, e.g., $0_{D(N)}$ designates the zero operator restricted to $D(N)$. Likewise

$$MT = \begin{pmatrix} N^* & 0 \\ 0 & N \end{pmatrix} \begin{pmatrix} 0 & 0 \\ A & 0 \end{pmatrix} = \begin{pmatrix} 0 & 0 \\ NA & 0 \end{pmatrix}.$$

Since $D(TM) = D(AN^*) \oplus D(N) \subset D(NA) \oplus L^2(\mathbb{R}) = D(MT)$, it ensues that $TM \subset MT$. Now, it is seen that

$$TM^* = \begin{pmatrix} 0 & 0 \\ A & 0 \end{pmatrix} \begin{pmatrix} N & 0 \\ 0 & N^* \end{pmatrix} = \begin{pmatrix} 0 & 0_{D(N^*)} \\ AN & 0 \end{pmatrix}$$

and

$$M^*T = \begin{pmatrix} N & 0 \\ 0 & N^* \end{pmatrix} \begin{pmatrix} 0 & 0 \\ A & 0 \end{pmatrix} = \begin{pmatrix} 0 & 0 \\ N^*A & 0 \end{pmatrix}.$$

Since $ANf \neq N^*Af$ for any $f \neq 0$, we infer that $TM^* \not\subset M^*T$ and $M^*T \not\subset TM^*$.

Remark Observe in the preceding examples that the crucial condition on domains is absent.

We have already observed that $BA \subset AB$ does not imply $BA^* \subset A^*B$ (or equivalently $B^*A \subset AB^*$), even when B is unitary, and A is closed and symmetric (Example 1.4.7). A similar question is:

$$BT \subset TB^* \Longrightarrow B^*T \subset TB$$

when $B \in B(H)$ is unitary, and T is closed and symmetric? The answer is still no, as seen next:

Example 5.4.3 Consider again a unitary $U \in B(H)$ and a closed (and symmetric) A such that $UA \subset AU$ and $U^*A \not\subset AU^*$, as in Example 1.4.7.
    Let

$$B = \begin{pmatrix} U & 0 \\ 0 & U^* \end{pmatrix} \text{ and } T = \begin{pmatrix} 0 & A \\ 0 & 0 \end{pmatrix}.$$

Then B is clearly unitary on $H \oplus H$ and T is closed on $H \oplus D(A)$. Moreover, the products BT and TB^* are well defined. More precisely,

$$BT = \begin{pmatrix} 0 & UA \\ 0 & 0 \end{pmatrix} \text{ and } TB^* = \begin{pmatrix} 0 & AU \\ 0 & 0 \end{pmatrix}$$

on the respective domains $D(BT) = D(T) = H \oplus D(A) = H \oplus D(UA)$ and $D(TB^*) = H \oplus D(AU)$. As $UA \subset AU$, it follows that $BT \subset TB^*$.

Now,

$$B^*T = \begin{pmatrix} 0 & U^*A \\ 0 & 0 \end{pmatrix} \text{ and } TB = \begin{pmatrix} 0 & AU^* \\ 0 & 0 \end{pmatrix}.$$

Since $U^*A \not\subset AU^*$, we plainly obtain $B^*T \not\subset TB$, as required.

If T is self-adjoint and if $B \in B(H)$, then $BT \subset TB$ do imply that $B^*T \subset TB^*$ (just take adjoints). One may therefore wonder whether $BT \subset TB^*$ implies $B^*T \subset TB$ in the case that T is self-adjoint and B is normal. This is again false as seen next:

Example 5.4.4 ([1]) The example to be given is a slight modification of the arguments used before. Consider a unitary $U \in B(H)$ and a closed A such that $UA \subset AU$ and $U^*A \not\subset AU^*$. Consider

$$B = \begin{pmatrix} U & 0 \\ 0 & U^* \end{pmatrix} \text{ and } T = \begin{pmatrix} 0 & A \\ A^* & 0 \end{pmatrix}.$$

Then B is unitary and T is closed on $D(A^*) \oplus D(A)$. Besides,

$$BT = \begin{pmatrix} 0 & UA \\ U^*A^* & 0 \end{pmatrix} \text{ and } TB^* = \begin{pmatrix} 0 & AU \\ A^*U^* & 0 \end{pmatrix}.$$

Since $UA \subset AU$, it results that $U^*A^* \subset A^*U^*$. Therefore, $BT \subset TB^*$. Since $U^*A \not\subset AU^*$ is equivalent to $UA^* \not\subset A^*U$, readers may easily check that

$$B^*T \not\subset TB$$

because $D(B^*T) = D(T) \not\subset D(TB)$, as wished.

References

1. I.F.Z. Bensaid, S. Dehimi, B. Fuglede, M.H. Mortad, The Fuglede theorem and some intertwining relations. Adv. Oper. Theory **6/1**, 8 pp. (2021). Paper No. 9
2. Ch. Chellali, M.H. Mortad, Commutativity up to a factor for bounded and unbounded operators. J. Math. Anal. Appl. **419/1**, 114–122 (2014)
3. S. Dehimi, M.H. Mortad, Unbounded operators having self-adjoint, subnormal or hyponormal powers. Math. Nachr. **54**, Art No. 7, 9pp. https://doi.org/10.13137/2464-8728/33883
4. S. Dehimi, M.H. Mortad, A. Bachir, Unbounded generalizations of the Fuglede-Putnam theorem. Rend. Inst. Mat. Univ. Trieste **54**(7), 9 (2022). https://doi.org/10.13137/2464-8728/33883
5. T. Furuta, On relaxation of normality in the Fuglede-Putnam theorem. Proc. Am. Math. Soc. **77**, 324–328 (1979)

6. Z.J. Jabłoński, I.B. Jung, J. Stochel, Unbounded quasinormal operators revisited. Integr. Equ. Oper. Theor. **79/1**, 135–149 (2014)
7. K.H. Jin, On unbounded subnormal operators. Bull. Korean Math. Soc. **30/1**, 65–70 (1993)
8. P.E.T. Jørgensen, Unbounded operators: perturbations and commutativity problems. J. Funct. Anal. **39/3**, 281–307 (1980)
9. M. Meziane, M.H. Mortad, Maximality of linear operators. Rend. Circ. Mat. Palermo, Ser II. **68/3**, 441–451 (2019)
10. M.H. Mortad, An application of the Putnam-Fuglede theorem to normal products of self-adjoint operators. Proc. Am. Math. Soc. **131/10**, 3135–3141 (2003)
11. M.H. Mortad, *Normal products of self-adjoint operators and self-adjointness of the perturbed wave operator on $L^2(\mathbb{R}^n)$*. Thesis (Ph.D.)-The University of Edinburgh (United Kingdom). ProQuest LLC, Ann Arbor, MI, 2003
12. M.H. Mortad, On some product of two unbounded self-adjoint operators. Integr. Equ. Oper. Theor. **64/3**, 399–408 (2009)
13. M.H. Mortad, An all-unbounded-operator version of the Fuglede-Putnam theorem. Complex Anal. Oper. Theor. **6/6**, 1269–1273 (2012)
14. M.H. Mortad. *An Operator Theory Problem Book* (World Scientific Publishing Co., Singapore, 2018)
15. M.H. Mortad, Yet another generalization of the Fuglede-Putnam theorem to unbounded operators. arXiv:2003.00339
16. A.E. Nussbaum, A commutativity theorem for unbounded operators in Hilbert space. Trans. Am. Math. Soc. **140**, 485–491 (1969)
17. S. Ôta, A quasi-affine transform of an unbounded operator. Studia Math. **112/3**, 279–284 (1995)
18. S. Ôta, K. Schmüdgen, On some classes of unbounded operators. Integr. Equ. Oper. Theor. **12/2**, 211–226 (1989)
19. F.C. Paliogiannis, A note on the Fuglede-Putnam theorem. Proc. Indian Acad. Sci. Math. Sci. **123/2**, 253–256 (2013)
20. F.C. Paliogiannis, A generalization of the Fuglede-Putnam theorem to unbounded operators. J. Oper. **2015**, 3 pp. (2015). Art. ID 804353
21. J.G. Stampfli, B.L. Wadhwa, An asymmetric Putnam-Fuglede theorem for dominant operators. Indiana Univ. Math. J. **25/4**, 359–365 (1976)
22. J. Stochel, An asymmetric Putnam-Fuglede theorem for unbounded operators. Proc. Am. Math. Soc. **129/8**, 2261–2271 (2001)
23. F.-H. Vasilescu, Anticommuting self-adjoint operators. Rev. Roumaine Math. Pures Appl. **28/1**, 76–91 (1983)

Chapter 6
Some Applications

In the beginning, why "some" applications only? The answer is pretty simple. The Fuglede theorem's applications are abundant and cannot all be included in this survey. Recall that the most useful application is that it weakens an assumption in the spectral theorem for normal operators.

6.1 Putnam's Applications

We give two results obtained by C.R. Putnam in [70].

Proposition 6.1.1 ([70], cf. [5]) *Any two similar normal operators are unitarily equivalent.*

Proof *([78], cf. [68])* Let T be an invertible operator such that $T^{-1}AT = B$, where A and B are normal. The aim is to show that there exists a unitary operator U such that $U^{-1}AU = B$.

Since T is invertible, it has a unique polar decomposition $T = UP$, where U is unitary and $P (= \sqrt{T^*T})$ is positive and invertible. Since $T^{-1}AT = B$, $TB = AT$, and since A and B are normal, the Fuglede-Putnam theorem gives $TB^* = A^*T$. Passing to adjoints yields

$$BT^* = (TB^*)^* = (A^*T)^* = T^*A.$$

Hence

$$BT^*T = T^*AT = T^*TB,$$

© The Author(s), under exclusive license to Springer Nature Switzerland AG 2022
M. H. Mortad, *The Fuglede-Putnam Theory*, Lecture Notes in Mathematics 2322,
https://doi.org/10.1007/978-3-031-17782-8_6

that is,

$$BP^2 = P^2B \text{ or } BP = PB$$

by standard arguments. Finally, we obtain

$$TB = AT \iff (UP)B = A(UP) \iff UBP = AUP \iff UB = AU.$$

Therefore, $U^{-1}AU = B$, and this completes the proof. $\square$

Remark Similarity was then weakened to just quasisimilarity in Lemma 4.1 in [21], and the authors of [66] then extended the latter to unbounded operators.

Theorem 6.1.2 ([70]) *Let A be an arbitrary operator, $C \in B(H)$ with $C \neq 0$, and let B be normal (non-necessarily bounded). Suppose that $AB \subset BA + C$ and $CB \subset BC$. Then A is unbounded.*

Proof *([70])* Assume that A is bounded, and let $B = \int z\,dE_z$ be the spectral resolution of B. Let Δ be any Borel set of the complex plane. Then, $E(\Delta) = \int_\Delta dE_z$. Since $CB \subset BC$, the Fuglede theorem yields $CE(\Delta) = E(\Delta)C$. Also, recall that by the spectral theorem, we know that $E(\Delta)B = BE(\Delta) = \int_\Delta z\,dE_z$.

Now, let Δ_1 and Δ_2 be two disjoint Borel sets of the complex plane. For $x \in D(AB) = D(B)$, we have $ABx = BAx + Cx$. Hence

$$E(\Delta_2)A \int_{\Delta_1} z\,dE_z x = \int_{\Delta_2} z\,dE_z AE(\Delta_1)x + E(\Delta_2)CE(\Delta_1)x.$$

As $CE(\Delta_1) = E(\Delta_1)C$, it ensues that

$$E(\Delta_2)CE(\Delta_1)x = E(\Delta_2)E(\Delta_1)C$$

for $E(\Delta_2)E(\Delta_1) = 0$. By an argument used by B. Fuglede (see [26]), we may obtain here too that $AE(\Delta) = E(\Delta)A$ for any Borel set in the complex plane. Therefore, $AB \subset BA$, whereby $Cx = 0$ for all $x \in D(B)$, i.e., $C \subset 0$, and since $D(B)$ is dense, it ensues that $C^* = 0$ and so $C = 0$, which is the sought contradiction. Consequently, A must be unbounded. $\square$

Remark The preceding result may be reformulated as follows: If B is normal, and $A, C \in B(H)$ are such that $AB \subset BA + C$ and $CB \subset BC$, then $C = 0$, thereby $AB \subset BA$.

Remark ([70]) Without the normality of B, Theorem 6.1.2 fails to hold even when all operators are in the form of 2×2 matrices. For instance, let

$$B = \begin{pmatrix} 1 & 1 \\ 0 & 1 \end{pmatrix}, \quad A = \begin{pmatrix} 2 & 0 \\ 0 & 1 \end{pmatrix} \text{ and } C = \begin{pmatrix} 0 & 1 \\ 0 & 0 \end{pmatrix}.$$

Then $AB + C = BA$ and $CB = BC$. Howbeit, $C \neq 0$, i.e., $AB \neq BA$, as wished.

6.2 Kaplansky's Theorem

At the time of [38], it was already known that if A, B, and AB are finite normal matrices, then BA is also normal. For a proof, see, e.g., Theorem 12.3.4 in [27]. I. Kaplansky gave a counterexample in [38] that shows the failure of the above result in an infinite-dimensional setting. However, he did show the following result:

Theorem 6.2.1 ([38]) *Let A and B be two bounded operators on a Hilbert space such that AB and A are normal. Then B commutes with AA^* iff BA is normal.*

Proof

1. "$\Rightarrow$": Since A is normal, it may be written as $A = UP$, where U is unitary, P is positive (recall that $P = \sqrt{A^*A} = |A|$) and P commutes with U. By hypothesis, B commutes with $AA^* = A^*A$ and so B commutes with P. Hence

$$U^*ABU = U^*UPBU = BPU = BA.$$

Therefore, BA is unitarily equivalent to AB, which is normal. Therefore, BA is normal too.

2. "$\Leftarrow$": We may write $A(BA) = (AB)A$. But both AB and BA are normal, the Fuglede-Putnam theorem then implies that

$$A(BA)^* = (AB)^*A \text{ or } AA^*B^* = B^*A^*A.$$

Taking adjoints yields $BAA^* = A^*AB$ which, by the normality of A, just means that B commutes with AA^*.

$\square$

Remark A similar paper is [28].

Let us restate a result from the previous proof separately:

Proposition 6.2.2 *If A and B are bounded operators on a Hilbert space H, A is normal and B commutes with A^*A, then the operators AB and BA are unitarily equivalent.*

Corollary 6.2.3 ([4]) *Let A and B be two bounded operators on a Hilbert space such that A is normal and AB is hyponormal. Then*

$$AA^*B = BAA^* \Longrightarrow BA \text{ is hyponormal.}$$

Another interesting consequence of the preceding theorem and T. Ando's theorem (see [3]) is the following result:

Corollary 6.2.4 ([4]) *Let A, B be two bounded operators such that A is normal. Assume that AB and $(BA)^*$ are paranormal. Assume further that the kernels of AB*

and $(AB)^*$ *(or those of* BA *and* $(BA)^*$*) coincide. Then*

$$AA^*B = BA^*A \Longleftrightarrow BA \text{ and } AB \text{ are normal.}$$

Remark T. Yamazaki and M. Yanagida showed in Corollary 3 in [91] that the paranormality of both T and T^* (where $T \in B(H)$) does make T normal, thereby improving Ando's result. In other words, the condition $\ker T = \ker T^*$ was superfluous, and at the time of [4], I was, unfortunately, unaware of this improvement (cf. [63]).

So, the revised version reads:

Proposition 6.2.5 *Let* A, B *be two bounded operators such that* A *is normal. Assume that* AB *and* $(BA)^*$ *are paranormal. Then*

$$AA^*B = BA^*A \Longleftrightarrow BA \text{ and } AB \text{ are normal.}$$

Next, we pass to generalizations of I. Kaplansky's theorem to unbounded operators.

Proposition 6.2.6 ([4]) *If* $B \in B(H)$ *is normal and* A *is a closed operator such that* BA *is normal and* $B^*BA \subset AB^*B$, *then*

1. AB is closed and densely defined, and $(AB)^*$ is formally normal.
2. AB is normal provided $D(AB) \subset D[(AB)^*]$

Remark Assuming that $D(AB) \subset D[(AB)^*]$ is weaker than the hyponormality of AB.

For the proof, we will need the following lemma:

Lemma 6.2.7 ([48], Proposition 1, [49] and [34], Lemma 2.1) *If* $BA \subset AB$ *where* $B \in B(H)$ *is self-adjoint and* A *is closed, then* $f(B)A \subset Af(B)$ *for any continuous function* f *on* $\sigma(B)$. *In particular, and if* B *is positive, then we have* $\sqrt{B}A \subset A\sqrt{B}$.

Remark In fact, this result was stated in [48] with the assumption "A being unbounded and self-adjoint", but by looking closely at its proof, we see that only the closedness of A was needed.

Let us provide a proof.

Proof Since $BA \subset AB$, it is seen that $p(B)A \subset Ap(B)$, for any real polynomial p. Now let f be a continuous function on $\sigma(B)$. Then, there exists a sequence (p_n) of polynomials converging uniformly to f so that

$$\lim_{n\to\infty} \|p_n(B) - f(B)\|_{B(H)} = 0.$$

Let $x \in D[f(B)A] = D(A)$. Let $t \in D(A)$ and $x = f(B)t$. Setting $x_n = p_n(B)t$, we see that

$$Ax_n = Ap_n(B)t = p_n(B)At \longrightarrow f(B)At.$$

Since $x_n \to x$, by the closedness of A, we get

$$x = f(B)t \in D(A) \text{ and } Ax = Af(B)t = f(B)At,$$

that is, we have proved that $f(B)A \subset Af(B)$. □

Now, we prove Proposition 6.2.6.

***Proof** ([4])*

1. We adapt the proof of I. Kaplansky's theorem [38] (with U unitary and $B = U|B| = |B|U$) to the context of unbounded operators. In the adaptation of Kaplansky's proof, we use Lemma 6.2.7, from which know that $|B|^2 A \subset A|B|^2$ implies $|B|A \subset A|B|$. Then we get $C \subset AB$, where $C = U^*BAU$ is normal. Since BA is densely defined, so is AB. Clearly AB is closed. Taking adjoints, we see that $(AB)^*$ is densely defined and $(AB)^* \subset C^*$. Since C^* is normal, $(AB)^*$ is formally normal.
2. Apply (1).

□

Corollary 6.2.8 ([4]) *Let A be a closed operator and let $B \in B(H)$ be such that both BA and B are normal. If AB is normal, then $B^*BA \subset AB^*B$.*

If AB is hyponormal and $B^*BA \subset AB^*B$, then AB is normal.

Proof We only prove the first assertion of the corollary. The proof in this case is a direct adaptation of I. Kaplansky's theorem. Since AB and BA are normal, the equation

$$B(AB) = (BA)B$$

implies that

$$B(AB)^* = (BA)^*B$$

by the Fuglede-Putnam theorem. Hence

$$BB^*A^* \subset A^*B^*B \text{ or } B^*BA \subset AB^*B.$$

□

6.3   Devinatz-Nussbaum's Theorem

We start by recalling a useful result of maximality of self-adjoint operators obtained by A. Devinatz et al. in [20]:

Theorem 6.3.1 *Let T, T_1 and T_2 be self-adjoint operators. Then*

$$T \subset T_1 T_2 \Longrightarrow T = T_1 T_2.$$

The aim is to show the following result about the strong commutativity of unbounded normal operators, which is a natural follow-up of Theorem 6.3.1.

Theorem 6.3.2 ([19], cf. Theorem 6.5.5) *If A, B and N are unbounded normal operators obeying $N = AB = BA$, then A and B strongly commute.*

A different proof was then obtained by A.E. Nussbaum in [65], and it is based upon Theorem 5.1.1. This proof requires the following proposition:

Proposition 6.3.3 *If A and B are normal operators in a Hilbert space H, and H is the orthogonal direct sum of subspaces M_n, $n = 1, 2, \cdots$, such that every M_n reduces A and B to normal operators A_n and B_n respectively, and if A_n commutes (strongly) with B_n for each n, then A and B are strongly commuting.*

Proof of Theorem 6.3.2 Nussbaum [65] Set $T = N^2$. So, T is normal and it commutes strongly with N. Besides, $D(T) \subset D(N) \subset D(A) = D(A^*)$ and $D(T) \subset D(AN)$. Hence

$$\langle Nx, A^*y \rangle = \langle ANx, y \rangle = \langle ABAx, y \rangle = \langle NAx, y \rangle = \langle Ax, N^*y \rangle$$

for all $x, y \in D(T)$. Theorem 5.1.1 entails the strong commutativity of N with A. Similarly, N strongly commutes with B.

So, there remains to show that A strongly commutes with B. Let E be the spectral measure associated with N, then set $E_n = E(\Delta)$ where $\Delta_n = \{z \in \mathbb{C} : n - 1 \leq |z| < n\}$. Then Δ_n commutes with A and B. Moreover, for all n

$$(AE_n)(BE_n) = ABE_n = BAE_n = (BE_n)(AE_n).$$

Thus, Proposition 6.3.3 does the remaining job, i.e., it implies the strong commutativity of A and B, as wished.   □

6.4 Embry's Thereom

The next result, which is due to M.R. Embry, even though its proof uses the Fuglede-Putnam theorem, is powerful enough to be compared with the Fuglede-Putnam theorem in some cases, as carried out in [54].

Theorem 6.4.1 ([23]) *Let $A, B \in B(H)$ be two commuting normal operators. Let $T \in B(H)$ be such that $0 \notin W(T)$. If $TA = BT$, then $A = B$.*

Proof *([23])* Let E and F be the spectral measures associated with A and B respectively. Since $TA = BT$, the Fuglede-Putnam theorem yields $TE(\Delta) = F(\Delta)T$ for any Borel set Δ of the complex plane. Since A and B commute, the Fuglede theorem implies that $E(\Delta)F(\Delta) = F(\Delta)E(\Delta)$, still for any Borel set Δ. Notice also that $T^*F(\Delta) = E(\Delta)T^*$, and that $T^2E(\Delta) = F(\Delta)T^2$. With these relations, readers may check that

$$[P(\Delta)]^*TP(\Delta) = [Q(\Delta)]^*TQ(\Delta) = 0, \tag{6.1}$$

where $P(\Delta) = [I - E(\Delta)]TE(\Delta)$ and $Q(\Delta) = E(\Delta)T[I - E(\Delta)]$ (where I denotes the identity operator on H). As $0 \notin W(T)$, Eq. (6.1) give $P(\Delta) = Q(\Delta) = 0$. More precisely, we have $TE(\Delta) = E(\Delta)T$ for any Borel set Δ. Therefore, $TA = AT$, whereby $AT = BT$. Hence $(A - B)T = 0$, and so

$$\langle (A - B)T(A - B)^*x, x \rangle = 0 \text{ or } \langle T(A - B)^*x, (A - B)^*x \rangle = 0$$

for all $x \in H$. So, if $A - B \neq 0$, then necessarily $0 \in W(T)$, which is flagrantly absurd. Accordingly, we must have $A = B$, as required. □

Remark The proof of M.R. Embry, although this is not mentioned in her paper, should work for two strongly commuting normal unbounded operators A and B, by keeping $T \in B(H)$, just as good. This generalization was obtained in [53] with an extra condition. However, and as will be seen below, an all-unbounded-operator version is not possible.

M.R. Embry obtained several interesting consequences of her theorem. Some of them are given below (see also Proposition 6.11.4).

Corollary 6.4.2 *Let $A \in B(H)$ be such that $0 \notin W(A)$. If AA^* and A^*A commute, then A is normal.*

Corollary 6.4.3 *Let $A \in B(H)$ be such that $0 \notin W(A)$. Assume that $B \in B(H)$. If $AB = -BA$, then $B = 0$, whenever A or B is normal.*

6.5 Normality of the Product of Normal Operators

We commence with a word on the case of self-adjoint operators. Let $A, B \in B(H)$ be self-adjoint. It is obvious that AB is self-adjoint if and only if $AB = BA$. We want to investigate this question further, and in two directions: Its generalization to unbounded self-adjoint operators; then normal operators (bounded or not).

Now, let A and B be two self-adjoint operators, where, this time, only $B \in B(H)$. Is AB self-adjoint? What about BA? As for BA, this is certainly not the case since an issue with closedness pops up. Indeed, if A is an unbounded self-adjoint that is also boundedly invertible with domain $D(A)$, and if $B = A^{-1} \in B(H)$, then BA is the identity operator restricted to $D(A)$, thereby BA is unclosed. For the case of AB, at least we do not have such a problem (indeed, AB is closed in this case). However, a naive direct attack may mislead some readers and make them believe that the best one can get is the symmetricity of BA. Indeed, by using "primitive tools" only, we obtain

$$BA \subset AB \iff BA \subset (BA)^{*},$$

i.e., BA is symmetric.

So, can we have a better outcome? The answer is affirmative, and the proof uses the notion of joint spectral measure.

Theorem 6.5.1 ([37] and [9]) *Let A and B be self-adjoint operators in H. Assume that $B \in B(H)$ and $BA \subset AB$. Then AB is a self-adjoint operator and $AB = \overline{BA}$. Moreover, if A and B are further assumed to be positive, then so is AB. In the end,*

$$\sqrt{AB} = \sqrt{A}\sqrt{B}.$$

Proof *([37] and [9])* By assumption and Theorem 6.3.2 in [7], the spectral measures E_A and E_B of A and B commute respectively. Let E be the joint spectral measure of the pair (A, B) (see Theorem 6.5.1 in [7]). Then, by Theorem 5.4.7 in [7], we have

$$\overline{BA} = \overline{AB} = \int_{\mathbb{R} \times \mathbb{R}} x \cdot t \, dE(x, t),$$

where $\mathbb{R}$ is the set of all real numbers. Since the operator AB is closed and the spectral integral appearing on the right-hand side of the foregoing displayed equation is self-adjoint (see Theorem 5.4.5 in [7]), we infer that AB is self-adjoint. To conclude, observe that $\overline{BA}$ too is self-adjoint. Since $BA \subset AB$, it ensues that $\overline{BA} \subset AB$. By maximality of self-adjoint operators, we obtain $AB = \overline{BA}$.

When A and B are positive, then we can replace $\mathbb{R} \times \mathbb{R}$ by $\mathbb{R}_+ \times \mathbb{R}_+$ in the spectral integral of $\overline{AB}$, where $\mathbb{R}_+ = [0, \infty)$. As a consequence, AB is positive.

Finally, we show the identity involving square roots. We may proceed as follows:

$$BA \subset AB \Longrightarrow \sqrt{B}A \subset A\sqrt{B} \Longrightarrow \sqrt{B}\sqrt{A} \subset \sqrt{A}\sqrt{B}$$

by Lemma 6.2.7 and Theorem 3.35 in [39].

By the first part of the proof, $\sqrt{A}\sqrt{B}$ is self-adjoint (and positive) and hence so is $(\sqrt{A}\sqrt{B})^2$. But,

$$(\sqrt{A}\sqrt{B})^2 = \sqrt{A}\sqrt{B}\sqrt{A}\sqrt{B} \subset \sqrt{A}\sqrt{A}\sqrt{B}\sqrt{B} = AB.$$

Next, as both $(\sqrt{A}\sqrt{B})^2$ and AB are self-adjoint, by maximality, we obtain

$$(\sqrt{A}\sqrt{B})^2 = AB.$$

Accordingly, and by uniqueness of the positive square root, we infer that

$$\sqrt{AB} = \sqrt{A}\sqrt{B},$$

as needed.   □

Remark The case of two unbounded self-adjoint operators as regards products will be merged with that of their normal counterparts. A related paper is [29].

Now, if $A, B \in B(H)$ are normal and $AB = BA$, then AB is normal. In the case of finite-dimensional spaces, there is a quite simple proof, which is "Fuglede-less", and it is based upon the so-called simultaneous unitary diagonalization. This technique is well-documented. Readers may consult it in Corollary 12.2.10 in [27].

The proof in infinite-dimensional spaces, and more generally the case of unbounded operators, seems to need the Fuglede theorem (or the spectral theorem).

Theorem 6.5.2 *Let $A, B \in B(H)$ be normal and such that $AB = BA$, then AB is normal.*

Proof Let A and B be two normal operators such that $AB = BA$, and so $B^*A^* = A^*B^*$. By the Fuglede theorem, we have $AB^* = B^*A$ and $A^*B = BA^*$ as well. Since A and B are normal, we have for all $x \in H$

$$\|ABx\| = \|A^*Bx\| = \|BA^*x\| = \|B^*A^*x\| = \|(AB)^*x\|,$$

i.e., AB is normal, as needed.   □

Remark The commutativity of A and B is indispensable as seen by considering for instance:

$$A = \begin{pmatrix} 1 & 0 \\ 0 & 2 \end{pmatrix} \text{ and } B = \begin{pmatrix} 0 & 1 \\ -1 & 0 \end{pmatrix}.$$

Then A is self-adjoint and B is normal. But

$$AB = \begin{pmatrix} 0 & -1 \\ 2 & 0 \end{pmatrix}$$

is non-normal.

Remark A related paper is [67].

A proof of the following result, which is borrowed from [46], was based on the Fuglede theorem but also on Theorem 6.3.1. Now, I present an adaptation of it, which bypasses Theorem 6.3.1, but it still requires Theorem 6.5.1. So, I don't see how one can avoid the spectral theorem (the consolation being that proofs of the Fuglede theorem for unbounded operators all use some results from the spectral theorem anyway!).

Theorem 6.5.3 *Let A, B be normal operators with $B \in B(H)$. If $BA \subset AB$, then AB and $\overline{BA}$ are both normal (and so $AB = \overline{BA}$).*

Proof *([46])* As $BA \subset AB$, $B^*A^* \subset A^*B^*$. By the Fuglede theorem, we have $BA^* \subset A^*B$ and $B^*A \subset AB^*$.

Let us show that AA^*BB^* (hence A^*AB^*B) is symmetric. We have

$$B^*A \subset AB^* \implies BB^*A \subset BAB^* \subset ABB^*$$

and hence

$$BB^*AA^* \subset ABB^*A^* \subset ABA^*B^* \subset AA^*BB^*.$$

Since both BB^* and AA^* are self-adjoint, and $BB^* \in B(H)$, we derive from Theorem 6.5.1 the self-adjointness of AA^*BB^*.

Now, $BA \subset AB$ yields $\overline{BA} \subset AB$. Hence

$$\overline{BA}(BA)^* \subset ABA^*B^* \subset AA^*BB^*.$$

By maximality, we obtain $\overline{BA}(BA)^* = AA^*BB^*$ as $\overline{BA}(BA)^*$ is also self-adjoint for $\overline{BA}$ is closed. In a similar way, we obtain

$$(BA)^*\overline{BA} \subset A^*AB^*B,$$

whereby $(BA)^*\overline{BA} = A^*AB^*B$, as above. Since A and B are normal, it is seen that

$$\overline{BA}(BA)^* = (BA)^*\overline{BA}.$$

Therefore, $\overline{BA}$ is normal.

To show that AB is normal, observe first that it is closed. Since $B^*A^* \subset A^*B^*$, the first part of the proof gives the normality of $\overline{B^*A^*}$. So, $(\overline{B^*A^*})^*$ is normal too. But

$$(\overline{B^*A^*})^* = (B^*A^*)^* = A^{**}B^{**} = \overline{A}B = AB.$$

In fine, since $\overline{BA}$ and AB are normal, the maximality of normal operators allows us to force $\overline{BA} = AB$ from $\overline{BA} \subset AB$.   □

Now, we give the general case of the normality (resp. self-adjointness) of two unbounded normal (resp. self-adjoint) operators.

Theorem 6.5.4 *Let A and B be two strongly commuting normal (resp. self-adjoint) operators. Then $\overline{AB} = \overline{BA}$ are normal (resp. self-adjoint).*

Proof *([9])* The proof makes use of the spectral theorem. Since A and B are normal, by the spectral theorem we may write

$$A = \int_{\mathbb{C}} z \, dE_A \text{ and } B = \int_{\mathbb{C}} z' \, dF_B,$$

where E_A and F_B designate the respective associated spectral measures. By the strong commutativity, we have

$$E_A(I)F_B(J) = F_B(J)E_A(I)$$

for all Borel sets I and J in $\mathbb{C}$. Hence

$$E_{A,B}(z, z') = E_A(z)F_B(z')$$

defines a two-parameter spectral measure. Thus

$$C = \int_{\mathbb{C}} \int_{\mathbb{C}} zz' \, dE_{A,B}$$

defines a normal operator, with $C = \overline{AB} = \overline{BA}$.

In the case of self-adjoint A and B, the same proof applies but all spectral integrals are over $\mathbb{R}$.   □

Remark If A and B are unbounded self-adjoint operators, then an expression like $AB = BA$ does not mean that A and B are strongly commuting. There is a famous example by E. Nelson in [64]. The first textbook probably to include Nelson's example is [75], and the same example is developed in detail in [80], pp. 257–258. Nevertheless, the simplest example perhaps is the one due to K. Schmüdgen (in [79], see also [83]). See also [62] for some more details.

The next result lies within the scope of this section. The proof, which utilizes the Fuglede theorem, may be found in [56].

Theorem 6.5.5 ([56], cf. [19]) *Let A and B be two unbounded normal operators verifying $AB \subset BA$. If B is boundedly invertible, then BA and $\overline{AB}$ are both normal whenever AB is densely defined. Furthermore, $\overline{AB} = BA$.*

In the end, we present a result involving hyponormal operators.

Theorem 6.5.6 ([9]) *Let A be a normal operator and let $B \in B(H)$ be hyponormal. If $BA \subset AB$, then A^*B is hyponormal, that is,*

$$\|(A^*B)^*x\| \le \|A^*Bx\|$$

for all $x \in D(A^*B) \subset D[(A^*B)^*]$.

By the Fuglede theorem, we then have:

Corollary 6.5.7 *Let A be a normal operator and let $B \in B(H)$ be hyponormal. If $BA \subset AB$, then AB is hyponormal.*

6.6 Normality of the Sum of Normal Operators

First, recall that if A and B are self-adjoint, with $B \in B(H)$, then $A + B$ is self-adjoint. This is seen by observing that

$$(A + B)^* = A^* + B^* = A + B.$$

Interested readers may find other results about the sum of self-adjoint unbounded operators in, e.g., [39, 57, 62, 76, 83].

Now, if $A, B \in B(H)$ are normal and $AB = BA$, then $A + B$ is normal. A "Fuglede-less" proof in finite-dimensional spaces is available in, e.g., Corollary 12.2.10 in [27].

As in the case of products, the proof in infinite-dimensional spaces (as well as the case of unbounded operators) seems to need the Fuglede theorem (or the spectral theorem). Let us supply proofs in this case.

Theorem 6.6.1 *Let $A, B \in B(H)$ be normal and such that $AB = BA$, then $A + B$ is normal.*

Proof We show that $A + B$ is normal. We have

$$(A + B)(A + B)^* = AA^* + AB^* + BA^* + BB^*.$$

Similarly,

$$(A + B)^*(A + B) = A^*A + A^*B + B^*A + B^*B.$$

Since A and B are normal, and since $AB^* + BA^* = A^*B + B^*A$, it follows that

$$(A + B)(A + B)^* = (A + B)^*(A + B),$$

which is a "synonym" of the normality of $A + B$.   □

Remark The commutativity of A and B is indispensable as seen by considering:

$$A = \begin{pmatrix} 1 & 0 \\ 0 & 2 \end{pmatrix} \text{ and } B = \begin{pmatrix} 0 & 1 \\ -1 & 0 \end{pmatrix}.$$

Then

$$A + B = \begin{pmatrix} 1 & 1 \\ -1 & 2 \end{pmatrix}$$

is not normal.

So, let us focus on the sum of two normal operators; one of them is unbounded, and before stating the result and giving its proof, we provide the following lemma (whose proof requires the Fuglede theorem):

Lemma 6.6.2 ([55]) *Let A be an unbounded normal operator with domain $D(A)$. Let $B \in B(H)$. If $BA \subset AB$, then*

1. $D(A^*(A + B)) = D(A^*A) \, (= D(A^*A) \cap D(A^*B))$;
2. $D(A(A^* + B^*)) = D(AA^*) \, (= D(AA^*) \cap D(AB^*))$.

The coming result appeared in [55].

Theorem 6.6.3 *Let A be an unbounded normal operator with domain $D(A)$. Let B be a bounded normal operator. Assume further that $BA \subset AB$. Then $A + B$ is normal on $D(A)$.*

Proof ([55]) First note that the closedness of $A + B$ is clear.

Since B is bounded and everywhere defined, $(A + B)^* = A^* + B^*$ and hence

$$(A + B)^*(A + B) = (A^* + B^*)(A + B) = A^*(A + B) + B^*(A + B).$$

While we always have $A^*(A + B) \supset A^*A + A^*B$, Lemma 6.6.2 implies $A^*(A + B) = A^*A + A^*B$.

Also since B is bounded, $B^*(A + B) = B^*A + B^*B$. Thus

$$(A + B)^*(A + B) = A^*A + A^*B + B^*A + B^*B. \tag{6.2}$$

Similarly

$$(A + B)(A + B)^* = A(A^* + B^*) + B(A^* + B^*).$$

Now let $x \in D(A^*A) = D(AA^*)$. We then have

$$
\begin{aligned}
\langle (A + B)x, (A + B)x \rangle &= \langle x, (A^*A + B^*A + A^*B + B^*B)x \rangle \\
&= \langle x, (AA^* + AB^* + BA^* + BB^*)x \rangle \\
&= \langle (A + B)^*x, (A + B)^*x \rangle.
\end{aligned}
$$

This leads to

$$\|(A + B)x\| = \|(A + B)^*x\| \text{ for all } x \in D(A^*A).$$

But $D(A^*A)$ is a core for A (see Theorem 3.24 in [39]) and for A^* and hence it is also a core for $A + B$ and for $(A + B)^*$. Thus $A + B$ is normal.   □

Let us now state the general result of two unbounded normal operators:

Theorem 6.6.4 *Let A and B be two strongly commuting unbounded normal operators. Then $\overline{A + B}$ is normal.*

Proof *([9])* Since A and B are normal, by the spectral theorem we may write

$$A = \int_{\mathbb{C}} z \, dE_A \text{ and } B = \int_{\mathbb{C}} z' \, dF_B,$$

where E_A and F_B denote the associated spectral measures. By the strong commutativity, we have

$$E_A(I)F_B(J) = F_B(J)E_A(I)$$

for all Borel sets I and J in $\mathbb{C}$. Hence

$$E_{A,B}(z, z') = E_A(z)F_B(z')$$

defines a two parameter spectral measure. Thus

$$C = \iint_{\mathbb{C}^2} (z + z') \, dE_{A,B}$$

defines a normal operator such that $C = \overline{A + B}$.   □

Remark Theorem 6.6.4 generalizes Theorem 6.6.3, as by taking $B \in B(H)$, it is seen that $A + B$ is closed, and the strong commutativity becomes equivalent to $BA \subset AB$ in such context.

Now, we present a result that gives the commutativity of two normal operators if their sum is normal.

Proposition 6.6.5 ([54], cf. [67]) *If A and B are two normal operators with $B \in B(H)$, and if AB^* and B^*A are self-adjoint, then the normality of $A + B$ implies that A and B commute.*

Proof Since A, B and $A + B$ are normal, we have

$$A^*B + B^*A = BA^* + AB^*.$$

The self-adjointness of AB^* and B^*A yields

$$BA^* \subset (AB^*)^* = AB^* \text{ and } (B^*A)^* = B^*A = A^*B.$$

Hence

$$B^*A \subset AB^*,$$

and the Fuglede theorem then yields the commutativity of A and B, as needed. $\square$

We close this section with a word on series of normal operators. This part is inspired by a known result that was shown in [72]. There, the writer established, under some conditions, the self-adjointness of the (weak) series $\sum_{n=1}^{\infty} G_n$ where (G_n) is a sequence of (strongly) commuting unbounded self-adjoint positive operators. Then, he inferred the self-adjointness of $\sum_{n=1}^{\infty} A_n^* A_n$, which represents the number of particles, where (A_n) is a countable family of unbounded operators having a dense domain. See pp. 88–91 in [72] for details and for what this series means in quantum mechanics (see [88] for the case of unbounded anti-commuting self-adjoint operators).

The analogous question in the case of normal operators has not been studied before, and it would constitute an interesting problem for researchers to push this question further. Here, we content ourselves to give a few simple consequences of some of the results above.

Theorem 6.6.6 *Let (B_n) be a sequence of pairwise commuting normal operators in $B(H)$. Then $\sum_{n=1}^{\infty} B_n$ is normal whenever this limit exists in norm.*

Proof By induction, it may be shown that a finite family of pairwise commuting normal operators is normal. So, $\sum_{n=1}^{k} B_n$ is normal. Since

$$\sum_{n=1}^{\infty} B_n = \lim_{k \to \infty} \sum_{n=1}^{k} B_n$$

is the uniform limit of a sequence of normal operators, it is itself normal, establishing the result.   □

It is known that the set of normal operators is not S.O.T.-closed (the weak limit of a sequence of normal operators need not remain normal either). See, e.g., [62] for counterexamples. However, we have:

Proposition 6.6.7 *Let (B_n) be a sequence of pairwise commuting normal operators in $B(H)$. Then $s - \sum_{n=1}^{\infty} B_n$ is subnormal whenever this limit exists.*

Proof As before, finite sums of pairwise commuting normal operators are normal. Taking the strong limit of those finite sums with a glance at Bishop's theorem ([8]) all yield the subnormality of $s - \sum_{n=1}^{\infty} B_n$.   □

6.7 Absolute Values of Products and Sums

In this section, we give several results involving the absolute value of the product and the sum of bounded and unbounded operators. Interested readers in such concepts may consult, e.g., [9, 30, 37, 58, 59] for further reading.

Proposition 6.7.1 *Let $A, B \in B(H)$ be such that $AB = BA$. If A is normal, then $|A||B| = |B||A|$.*

Remark The previous result was shown in [93] by assuming that both A and B are normal.

Proof *([59])* Since $AB = BA$ and A is normal, we have $A^*B = BA^*$. We then have from the previous two relations:

$$AB = BA \implies A^*AB = A^*BA = BA^*A.$$

Hence

$$|A|B = B|A|.$$

Since $|A|$ is self-adjoint, the previous equality gives (by taking adjoints) $|A|B^* = B^*|A|$. Hence

$$B^*|A|B = |A|B^*B \implies B^*B|A| = |A|B^*B \implies |B||A| = |A||B|,$$

as required.   □

It was noted in [32] that if S, T are two non-commuting self-adjoint operators, then the inequality $|ST| \leq |S||T|$ never holds. So, the next result gives a condition of when such an inequality holds:

Theorem 6.7.2 ([59]) *Let $A, B \in B(H)$ be self-adjoint such that AB is normal. Then*

$$|AB| = |A||B|.$$

Proof *([59])* Since A and B are self-adjoint, we may write

$$B(AB) = BAB = (AB)^*B.$$

Since AB and $(AB)^*$ are normal, the Fuglede-Putnam theorem gives

$$B(AB)^* = (AB)^{**}B \text{ or merely } B^2A = AB^2.$$

Consequently, $B^2A^2 = AB^2A = A^2B^2$.

On the other hand, we easily see that

$$|AB|^2 = (AB)^*AB = AB(AB)^* = AB^2A = A^2B^2$$

and so

$$|AB| = \sqrt{A^2B^2} = \sqrt{A^2}\sqrt{B^2} = |A||B|,$$

as required. $\square$

Since $|AB|$ is self-adjoint, we have:

Corollary 6.7.3 *Let $A, B \in B(H)$ be self-adjoint such that AB is normal. Then $|A||B|$ is self-adjoint, i.e., $|A||B| = |B||A|$.*

Now, we pass to the case of unbounded operators.

Proposition 6.7.4 *Let A and B be two strongly commuting normal operators. Then*

$$\overline{|BA|} = \overline{|AB|} = \overline{|A||B|} = \overline{|B||A|}.$$

Proof *([9])* We have seen in Theorem 6.5.4 (by keeping the same notations) that

$$C = \int_{\mathbb{C}} \int_{\mathbb{C}} zz' dE_{A,B}$$

defines a normal operator, such that $C = \overline{AB} = \overline{BA}$. Therefore, as $|zz'| = |z||z'|$ for all z, z', then (see, e.g., p. 78 in [83])

$$|\overline{AB}| = \int_{\mathbb{R}} \int_{\mathbb{R}} |zz'| dE_{A,B} = \overline{|A||B|},$$

as wished.   □

Proposition 6.7.5 *Let A and C be strongly commuting normal operators and let $B \in B(H)$. If $BA \subset AB$, $BC \subset CB$ and AC is densely defined, then*

$$|\overline{ABC}| = |\overline{AC}||B| = \overline{|A||C|}|B|.$$

The proof of the preceding proposition relies upon Lemma 1.2.5.

Proof ([9]) By assumptions,

$$B(AC) \subset ABC \subset (AC)B.$$

Since AC is densely defined and $D(AC) = D(BAC) \subset D(ABC)$, clearly ABC is densely defined. Hence

$$(ABC)^* \subset [B(AC)]^* = (AC)^* B^*.$$

As $\overline{AC}$ is normal, Lemma 1.2.5 gives

$$B(AC)^* \subset (AC)^* B \text{ or } B^* \overline{AC} \subset \overline{AC} B^*.$$

Also,

$$ABC \subset ACB \subset \overline{AC}B$$

and so

$$\overline{ABC} \subset \overline{AC}B$$

for $\overline{AC}B$ is closed. Hence, we may write

$$(ABC)^* \overline{ABC} \subset (AC)^* B^* \overline{ABC} \subset (AC)^* B^* \overline{AC}B \subset (AC)^* \overline{AC}B^* B.$$

As in the proof of Theorem 6.5.1, we may obtain the self-adjointness of $(AC)^* \overline{AC} B^* B$, and because $\overline{ABC}$ is closed, we then get

$$(ABC)^* \overline{ABC} = (AC)^* \overline{AC}B^* B.$$

Therefore, Theorem 6.5.1 yields

$$\overline{|ABC|} = \overline{|AC|}|B| = \overline{|A|}|C||B|,$$

establishing the result. □

Now, we pass to the case of inequalities involving bounded operators. It is well-known that the inequality

$$|A + B| \leq |A| + |B|$$

need not even hold for two 2×2 self-adjoint matrices. See, e.g., [62] for a counterexample. Adding a commutativity condition on A and B (e.g., both of them normal) is sufficient for that inequality to hold.

The main result that does not need any advanced spectral tools reads:

Theorem 6.7.6 *Let $A, B \in B(H)$ be such that $AB = BA$. If A is normal and B is hyponormal, then the following triangle inequality holds:*

$$|A + B| \leq |A| + |B|.$$

The proof needs a couple of fairly standard auxiliary lemmata.

Lemma 6.7.7 *Let $T \in B(H)$ be hyponormal (i.e., $TT^* \leq T^*T$, that is, $\|T^*x\| \leq \|Tx\|$ for all $x \in H$). Then*

$$\operatorname{Re} T = \frac{T + T^*}{2} \leq \sqrt{T^*T} = |T|.$$

Lemma 6.7.8 *Let $A, B \in B(H)$ be such that A is normal and B is hyponormal. If $AB = BA$, then A^*B is hyponormal.*

We are ready to show Theorem 6.7.6.

Proof *([59])* Since A is normal and $AB = BA$, we know from Proposition 6.7.1 that $|A||B| = |B||A|$. Hence

$$|A + B|^2 \leq (|A| + |B|)^2 \iff (A + B)^*(A + B) \leq |A|^2 + |B|^2 + 2|A||B|$$

$$\iff A^*B + B^*A \leq 2\sqrt{A^*A}\sqrt{B^*B}.$$

We already know from above that $\sqrt{A^*A}\sqrt{B^*B} = \sqrt{A^*AB^*B}$. So, to prove the desired triangle inequality, we are only required to prove

$$A^*B + B^*A \leq 2\sqrt{A^*AB^*B}.$$

But

$$A^*AB^*B = AA^*B^*B = AB^*A^*B = B^*AA^*B.$$

If we set $T = A^*B$, then are done with the proof if we come to show that $T + T^* \le 2\sqrt{T^*T}$ holds. But this is just Lemma 6.7.7 once we show that A^*B is hyponormal. This is in effect the case as A^*B is hyponormal by Lemma 6.7.8.

Therefore, we have shown that

$$|A + B|^2 \le (|A| + |B|)^2$$

and so (by Heinz inequality, say)

$$|A + B| \le |A| + |B|,$$

as needed.   □

In the case of inequalities involving unbounded operators, one has to be extremely careful with the order relation to be used. We give two definitions that coincide in the context of bounded operators. See [69] for more about this topic.

Definition 6.7.1 (Definition 10.5, p. 230 in [83]) Let A and B be two symmetric operators with domains $D(A)$ and $D(B)$ respectively. We write $A \succeq B$ if $D(A) \subset D(B)$ and

$$\langle Ax, x \rangle \ge \langle Bx, x \rangle, \ \forall x \in D(A).$$

Definition 6.7.2 Let T and S be unbounded positive self-adjoint operators. Say that $S \ge T$ provided that $D(S^{1/2}) \subset D(T^{1/2})$ and $\left\| S^{1/2}x \right\| \ge \left\| T^{1/2}x \right\|$ for all $x \in D(S^{1/2})$.

Let us recall a couple of useful results. The Heinz inequality is valid for positive unbounded self-adjoint operators (see, e.g., [83]). So, if S and T are self-adjoint operators, then

$$S \ge T \ge 0 \Longrightarrow \sqrt{S} \ge \sqrt{T}.$$

Also, if S and T are self-adjoint, and T is boundedly invertible, then $S \ge T \ge 0$ implies that S is boundedly invertible and $S^{-1} \le T^{-1}$ (see, e.g., p. 200 in [89]; cf. [16] and [60]).

The next result cannot be shown using the spectral theorem as one of the operators is hyponormal (it also generalizes Theorem 6.7.6).

Theorem 6.7.9 *Let A be a normal operator and let $B \in B(H)$ be hyponormal. If $BA \subset AB$, then*

$$|A + B| \leq |A| + |B|.$$

Proof *([9])* First, as usual we have $BA \subset AB$ and $BA^* \subset A^*B$. Whence, if $x \in D(A) = D(A^*)$, then $Bx \in D(A) = D(A^*)$. Moreover, the self-adjointness of $|A|$ and $|B|$ gives the self-adjointness (and positiveness) of $|A| + |B|$ for $|B| \in B(H)$. Since A is closed and $B \in B(H)$, $A + B$ is closed. Hence $|A + B|$ is well-defined, and besides

$$D(|A| + |B|) = D(|A|) = D(A) = D(A + B) = D(|A + B|).$$

Now, since $|A + B|$ is self-adjoint and positive, to show the required triangle inequality, by calling on the Heinz inequality, it suffices to show that

$$|A + B|^2 \leq (|A| + |B|)^2.$$

Let $x \in D(A)$. Then

$$\||A + B|x\|^2 = \|(A + B)x\|^2$$
$$= \langle (A + B)x, (A + B)x \rangle$$
$$= \|Ax\|^2 + \|Bx\|^2 + \langle Bx, Ax \rangle + \langle Ax, Bx \rangle$$
$$= \|Ax\|^2 + \|Bx\|^2 + \langle Bx, Ax \rangle + \overline{\langle Bx, Ax \rangle}$$
$$= \|Ax\|^2 + \|Bx\|^2 + 2\operatorname{Re}\langle Bx, Ax \rangle$$
$$= \|Ax\|^2 + \|Bx\|^2 + 2\operatorname{Re}\langle A^*Bx, x \rangle$$
$$\leq \|Ax\|^2 + \|Bx\|^2 + 2|\langle A^*Bx, x \rangle|.$$

But A^*B is closed. It is also hyponormal by Theorem 6.5.6 and so Theorem 3.2 in [16] yields

$$|\langle A^*Bx, x \rangle| \leq \langle |A^*B|x, x \rangle$$

for each x. By the results above, we have for each $x \in D(A)$:

$$|A^*B|x = |A^*||B|x = |A||B|x = |B||A|x.$$

Hence

$$2|\langle A^*Bx, x \rangle| \leq 2\langle |A^*B|x, x \rangle = \langle |A||B|x, x \rangle + \langle |B||A|x, x \rangle.$$

Accordingly,

$$\||A + B|x\|^2 \leq \|Ax\|^2 + \|Bx\|^2 + \langle |A||B|x, x\rangle + \langle |B||A|x, x\rangle = \|(|A| + |B|)x\|^2$$

or merely

$$\||A + B|x\| \leq \|(|A| + |B|)x\|$$

for all $x \in D(A)$. By Definition 6.7.2, this just means that

$$|A + B|^2 \leq (|A| + |B|)^2,$$

as needed above.   □

We close this section by the general case of two unbounded normal operators.

Theorem 6.7.10 ([9]) *Let A and B be two strongly commuting unbounded normal operators. Then,*

$$|\overline{A + B}| \leq |A| + |B|.$$

Proof *([9])* Since A and B are normal, by the spectral theorem we may write

$$A = \int_{\mathbb{C}} z \, dE_A \text{ and } B = \int_{\mathbb{C}} z' \, dF_B,$$

where E_A and F_B denote the associated spectral measures. As A and B strongly commute, then so do $|A|$ and $|B|$. Since $|A|$ and $|B|$ are also self-adjoint and positive, by Lemma 4.15.1 in [72], it follows that $|A| + |B|$ is self-adjoint (hence closed) and positive. As for domains, we have $D(|A|+|B|) = D(|A|) \cap D(|B|) = D(A) \cap D(B)$ and $D(A + B) \subset D(\overline{A + B}) = D(|\overline{A + B}|)$.

By the strong commutativity, we have

$$E_A(I)F_B(J) = F_B(J)E_A(I)$$

for all Borel sets I and J in $\mathbb{C}$. Hence

$$E_{A,B}(z, z') = E_A(z)F_B(z')$$

defines a two-parameter spectral measure. Thus

$$C = \iint_{\mathbb{C}^2} (z + z') \, dE_{A,B}$$

defines a normal operator such that $C = \overline{A + B}$. Since $|z + z'| \leq |z| + |z'|$ for all z, z', it follows that

$$|\overline{A + B}| = \iint\limits_{\mathbb{R}^2} |z + z'| dE_{A,B}(z, z') \leq |A| + |B|$$

(we used the positivity of the measure $\langle E_{A,B}(\Delta)x, x \rangle$ with Δ being a Borel subset of $\mathbb{R}^2$). In fact, what we have proved so far is "only"

$$|\overline{A + B}| \preceq |A| + |B|.$$

In other language, we have shown that

$$\langle |\overline{A + B}|x, x \rangle \leq \langle (|A| + |B|)x, x \rangle$$

for all $x \in D(|A| + |B|) \subset D(|\overline{A + B}|)$.

Since $|\overline{A + B}|$ and $|A| + |B|$ are self-adjoint and positive, by Lemma 10.10 in [83], "$\preceq$" becomes "$\leq$", that is, we have established the desired inequality:

$$|\overline{A + B}| \leq |A| + |B|.$$

$\square$

6.8   Self-Adjointness of the Normal Product of Self-Adjoint Operators

We already know that if $A, B \in B(H)$ are self-adjoint, then AB is self-adjoint if and only if $AB = BA$. Under what conditions is a product of two self-adjoint operators self-adjoint whenever it is normal? It appears that W. Rehder was the first mathematician to investigate this question in his paper [77]. Unfortunately, E. Albrecht et al. were all unaware of Rehder's result at the time of their respective papers [1], and [48]. We state all of these results in the present section.

Theorem 6.8.1 ([77]) *If $A, B \in B(H)$ are self-adjoint such that one of them is positive, then AB is self-adjoint whenever AB is normal.*

Proof Suppose A is positive. Since both AB and $(AB)^*$ ($=BA$) are normal, the Fuglede-Putnam theorem gives

$$A(BA) = (AB)A \iff A(AB)^* = (AB)A \iff A(AB)^{**} = (AB)^*A,$$

i.e., $A^2B = BA^2$ or $AB = BA$ (as $A \geq 0$). Therefore, AB is self-adjoint.

The proof when $B \geq 0$ is identical by using the relation $B(AB) = (BA)B$, hence it is omitted.   □

Remark Another proof appeared in [1]. Let $A, B \in B(H)$ be self-adjoint such that $A \geq 0$, say. Then as in [33], we know that

$$\sigma(AB) = \sigma(BA) = \sigma(\sqrt{A}B\sqrt{A}).$$

Since $\sqrt{A}B\sqrt{A}$ is self-adjoint, it has a real spectrum, and so AB too has a real spectrum. Since a normal operator with a real spectrum is self-adjoint, we infer that AB is self-adjoint.

E. Albrecht and P.G. Spain obtained the following result as a consequence of another result established in complex unital Banach algebras.

Theorem 6.8.2 ([1]) *Let $A, B \in B(H)$ be self-adjoint and such that $\sigma(A) \cap \sigma(-A) \subseteq \{0\}$. If AB is normal, then it is self-adjoint.*

Remark Notice that a condition like $\sigma(A) \cap \sigma(-A) \subseteq \{0\}$ is essentially stronger than $A \geq 0$, and a positive A need not satisfy such an asymmetric condition. See [62] for complementary counterexamples.

The following result generalizes Theorem 6.8.1 (cf. [15]).

Proposition 6.8.3 *Let $A, B \in B(H)$ be both self-adjoint. If BA is hyponormal and B is positive, or if AB is hyponormal and A is positive, then AB is self-adjoint, that is, $AB = BA$.*

Proof Write

$$BAB = B(BA)^* = (BA)B.$$

If BA is hyponormal, it follows by Theorem 2.3.2 that $B(BA) = (BA)^*B$ or merely $B^2A = AB^2$. Since $B \geq 0$, we infer $BA = AB$, establishing the self-adjointness of AB.

A similar reasoning applies in the case of the hyponormality of AB. Write $A(AB)^* = (AB)A$. Theorem 2.3.2 then yields $A^2B = BA^2$, and so $AB = BA$ when $A \geq 0$, i.e., we have shown that AB is self-adjoint.   □

Remark Notice in the end that the previous corollary improves a similar result obtained in [36]. There, the author showed the same result by assuming, in addition to the above hypotheses, that B had a dense range.

Proposition 6.8.3 may easily be extended to the following result using the same idea of its proof.

Proposition 6.8.4 *Let $A, B \in B(H)$ be both self-adjoint. If BA is p-hyponormal and B is positive, or if AB is p-hyponormal and A is positive, then AB is self-adjoint, that is, $AB = BA$.*

Before turning to unbounded operators, we stop by some similar results in the bounded case. Obviously, a unitary operator is an isometry. Conversely, a self-adjoint (or normal) isometry is unitary. So, the next result might have a particular interest to some readers.

Proposition 6.8.5 ([51]) *Assume that A and B are two self-adjoint operators such that say $\sigma(B) \cap \sigma(-B) \subseteq \{0\}$. If AB is an isometry, then it is self-adjoint.*

Proof *([51])* Set $N = AB$. Since A and B are self-adjoint,

$$BN = BAB = (BA)B = N^*B.$$

Now as N is an isometry, Proposition 2.1.3 gives us

$$BN^* = NB, \text{ i.e., } B^2A = AB^2.$$

Since $\sigma(B) \cap \sigma(-B) \subseteq \{0\}$, we obtain

$$BA = AB, \text{ i.e., } N = N^*$$

completing the proof. □

Corollary 6.8.6 *Let A be a projection onto a closed subspace $\mathcal{M}$. If B is self-adjoint, then*

$$AB \text{ isometry} \implies B \text{ reduces } \mathcal{M}.$$

Next, we treat the case of one unbounded operator.

Theorem 6.8.7 ([48]) *Let A and B be two self-adjoint operators where $B \in B(H)$. Assume further that B is positive and that AB is normal. Then AB is self-adjoint, and $AB = \overline{BA}$.*

Proof *([48])* Write

$$BAB = B(AB) = (BA)B \subset (AB)^*B.$$

Since AB is normal, so is $(AB)^*$. By the Fuglede-Putnam theorem, we get

$$B^2A = BBA \subset B(AB)^* \subset (AB)^{**}B = \overline{AB}B = ABB = AB^2.$$

Since B is positive, we obtain $BA \subset AB$. Hence

$$(AB)^* \subset (BA)^* = AB,$$

and because $D(AB) = D[(AB)^*]$, we finally derive $AB = (AB)^*$, i.e., AB is self-adjoint. □

Corollary 6.8.8 *Let A and B be two self-adjoint operators, where $B \in B(H)$. Assume further that B is positive and that BA is normal. Then BA is self-adjoint, and $AB = BA$.*

Proof Since BA is normal, it ensues that $(BA)^* = AB$ too is normal. By the preceding theorem, AB or $(BA)^*$ must be self-adjoint. In another language,

$$(BA)^* = (BA)^{**} = \overline{BA} = BA,$$

i.e., we have reached the self-adjointness of BA, as needed.   □

The next result, which was obtained a while after that in [48], deals with the case of when the unbounded operator A is positive whilst B is self-adjoint and bounded.

Theorem 6.8.9 ([30]) *Let A and B be two self-adjoint operators, where $B \in B(H)$. Assume further that A is positive and that BA is normal. Then both BA and AB are self-adjoint, and $AB = BA$.*

Proof *([30])* We may write

$$A(BA) = (AB)A = (BA)^*A.$$

Since BA is normal, $(BA)^*$ is normal too. Since $D(BA) = D(A)$, Theorem 5.1.2 applies and yields

$$A(BA)^* \subset BAA$$

or

$$A^2B \subset BA^2. \tag{6.3}$$

Let us transform the previous into a commutativity between B and A^2 (i.e., $BA^2 \subset A^2B$).

Since BA and $(BA)^*$ are normal, we may write

$$B(BA)^* = B(AB) = (BA)B \implies B(BA) = (BA)^*B$$

or

$$B^2A = AB^2. \tag{6.4}$$

This tells us that both B^2A and AB^2 are self-adjoint. Continuing we note that

$$B^2A^2 = AB^2A = A^2B^2$$

and

$$B^2A^4 = B^2A^2A^2 = A^2B^2A^2 = A^4B^2. \tag{6.5}$$

To prove B commutes with A^2, we first show that $\overline{BA^2}$ is normal. We have

$$
\begin{aligned}
(BA^2)^*BA^2 &= A^2BBA^2 \\
&\supset A^2BA^2B \text{ (by Inclusion (6.3))} \\
&\supset A^2A^2BB \text{ (by Inclusion (6.3))} \\
&= A^4B^2 \\
&= B^2A^4.
\end{aligned}
$$

Passing to adjoints gives

$$(\overline{BA^2})^*\overline{BA^2} = (BA^2)^*\overline{BA^2} \subset [(BA^2)^*BA^2]^* \subset (B^2A^4)^* = A^4B^2.$$

But A^4B^2 is self-adjoint. Since $\overline{BA^2}$ is closed, $(\overline{BA^2})^*\overline{BA^2}$ is self-adjoint, and since self-adjoint operators are maximally symmetric, we immediately obtain

$$(\overline{BA^2})^*\overline{BA^2} = A^4B^2. \tag{6.6}$$

Similarly, we may obtain

$$B^2A^4 = A^4B^2 = A^2A^2BB \subset A^2BA^2B \subset BA^2A^2B = BA^2(BA^2)^*$$

and passing to adjoints yields

$$\overline{BA^2}(\overline{BA^2})^* = \overline{BA^2}(BA^2)^* \subset [BA^2(BA^2)^*]^* \subset (B^2A^4)^* = A^4B^2.$$

Similar arguments as above imply that

$$\overline{BA^2}(\overline{BA^2})^* = A^4B^2. \tag{6.7}$$

By Eqs. (6.6) and (6.7), we see that $\overline{BA^2}$ is normal and hence we deduce as

$$(\overline{BA^2})^* = (BA^2)^* = A^2B$$

that A^2B is closed, in fact normal.

Since $A^2 B$ is densely defined, we may adjoint Relation (6.3) to obtain

$$(A^2 B)^* \supset (BA^2)^* = A^2 B$$

from which $A^2 B$ is symmetric. Since we have just seen that $A^2 B$ is normal, we infer that $A^2 B$ is self-adjoint. Thus, we have arrived at the basic inclusion and commutativity relation

$$BA^2 \subset \overline{BA^2} = (A^2 B)^* = A^2 B = (BA^2)^*.$$

In particular, we then know from Theorem 10 in [6] (or [39]) and the positivity of A that B commutes with A, that is,

$$BA \subset AB \, (= (BA)^*).$$

But both BA and $(BA)^*$ are normal. Since normal operators are maximally normal, we obtain $BA = AB$.

Accordingly,

$$BA = AB = (BA)^* = (AB)^*,$$

and this completes the proof.   □

Now, we turn to the case of AB normal (keeping all the other assumptions, except BA normal, as those of Theorem 6.8.9). It took some time before a breakthrough was reached. The main reason was that the version of the Fuglede theorem that could give the result was untrue (see [30] for more details). Thus the method of proof of Theorem 6.8.9 could not be applied to the case AB, and the approach had to be different.

We state the following result without proof. Notice that the proof requires several lemmata that are also interesting in their own right.

Theorem 6.8.10 ([30]) *Let A and B be two self-adjoint operators where only B is bounded. Assume further that A is positive and that AB is normal. Then both $\overline{BA}$ and AB are self-adjoint. Besides one has $AB = \overline{BA}$.*

There is still a case worth investigating, namely: The normality of the product of *two unbounded* self-adjoint operators; one of them is positive. The answer depends on which of the factors in the product is positive.

Theorem 6.8.11 ([48]) *Let A and B be two unbounded self-adjoint operators such that $B \geq 0$. If AB is normal, then it is self-adjoint.*

Proof *([48])* Set $N = AB$. We have

$$BAB = B(AB) = (BA)B \subset (AB)^* B$$

which implies that $BN \subset N^*B$. But $D(N) \subset D(B)$, so by Theorem 5.1.2, $BN^* \subset NB$ or

$$B^2A \subset B(AB)^* \subset AB^2.$$

In other words, $B^2Ax = AB^2x$ for $x \in D(B^2A)$. Reasoning as in the proof of Theorem 5.1.2, we may say that for $x \in \text{ran } P_{B_n}$ we have $B^2ABx = AB^2Bx$ as $Bx \in D(B^2A)$ since B^2N is bounded in this case. We have $B^2Nx = NB^2x$. Hence $BNx = NBx$. But $BNx = N^*Bx$ on H_n. Whence $N^*Bx = NBx$.

We now use the orthogonal decomposition $H_n = \overline{\text{ran } B} \oplus \ker B$ for the B restricted to H_n. So, $N = N^*$ on $\overline{\text{ran } B}$, and both are 0 on $\ker B$. Hence $N = N^*$ on H_n. This shows that N_n (where N_n designates N restricted to H_n) is self-adjoint. Hence $\sigma(N_n) \subset \mathbb{R}$ for all n, and then $\sigma(N) \subset \mathbb{R}$. Remembering that a normal operator with a real spectrum is automatically self-adjoint, we finally infer that AB is self-adjoint, as coveted.   □

Remark The case where the left factor of the product is positive is false. See Example 6.14.3 for a counterexample.

The question of the essential self-adjointness of a product of two self-adjoint operators is arduous. In [48], a three-page counterexample was constructed to show that if A and B are two unbounded self-adjoint operators such that $B \geq 0$, then the normality of $\overline{AB}$ does not entail its self-adjointness (a non-detailed version of it is Example 6.14.4). Related to the question of essential self-adjointness of products, the reader may consult [47]. That being said, we may rephrase the result of Theorem 6.8.10 as follows:

Corollary 6.8.12 *Let A and B be two self-adjoint operators where $B \in B(H)$. Assume further that A is positive and that $\overline{BA}$ is normal. Then BA is essentially self-adjoint.*

Remark Most of the results above were improved in [18].

A generalization of Corollary 6.8.8 is given next. The proof is not based upon the Fuglede-Putnam theorem, but below it we say a word of how to obtain the same result using one of the versions of the Fuglede-Putnam theorem.

Theorem 6.8.13 ([15]) *Let A and B be two self-adjoint operators such that B is bounded and positive. If BA is hyponormal, then both BA and AB are self-adjoint (and $AB = BA$!) whenever $\sigma(BA) \neq \mathbb{C}$.*

The proof is based on the next proposition:

Proposition 6.8.14 ([15]) *Let $P \in B(H)$ be self-adjoint and let A be an arbitrary operator such that P^2A is closed. Then PA is closed.*

Now, we prove Theorem 6.8.13.

Proof *([15])* Let P be the unique square root of B. Since $\sigma(BA) \neq \mathbb{C}$, BA or P^2A is closed so that PA is closed by Proposition 6.8.14. The rest of the proof is divided into two parts.

1. First, PAP is self-adjoint: Since P is bounded and PA is closed, we have

$$(PAP)^* = (AP)^*P^* = (AP)^*P = (PA)^{**}P = \overline{PA}P = PAP,$$

   i.e., PAP is surely self-adjoint so that $\sigma(PAP) \neq \mathbb{C}$.
2. Second, we show that BA and AB are self-adjoint: Since $\sigma(P^2A) \neq \mathbb{C}$ and $\sigma(PAP) \neq \mathbb{C}$, by Corollary 3.4 in [15]

$$\sigma(BA) = \sigma(PAP) \subset \mathbb{R}.$$

Now, if $W(BA)$ denotes the numerical range of BA, then from [35] we know that

$$W(BA) \subset \operatorname{conv} \sigma(BA) \subset \mathbb{R}$$

for BA is hyponormal. Thus BA is closed, symmetric and with real spectrum, it is self-adjoint! Accordingly,

$$AB = (BA)^* = BA.$$

□

Remark If we assume that BA is subnormal (which is *stronger* than hyponormal), then we can obtain the self-adjointness of BA and AB without using the machinery of the preceding proof, we merely apply Theorem 5.3.5.

Since the spectrum of an everywhere defined bounded operator is compact (hence necessarily different from $\mathbb{C}$), we have an immediate consequence (cf. Proposition 6.8.3):

Corollary 6.8.15 *Let $A, B \in B(H)$ be two self-adjoint operators such that B is positive. If BA is hyponormal, then both BA and AB are self-adjoint, that is, $AB = BA$.*

Remark We note in the end that the similar question in the case of the sum of two *unbounded* operators is obvious. Indeed, if A and B are two unbounded self-adjoint operators such that $A + B$ is normal, then $A + B$ is automatically self-adjoint. Indeed, since A and B are self-adjoint,

$$A + B = A^* + B^* \subset (A + B)^*,$$

i.e., $A + B$ is symmetric. Since it is also normal, it follows that $A + B$ is self-adjoint.

6.8.1 Commutativity Up to a Factor

Definition 6.8.1 We say $A, B \in B(H)$ commute up to a factor when $AB = \lambda BA \neq 0$, $\lambda \in \mathbb{C}^*$ (we may also say that A and B are λ-commuting).

Remark The relation $AB = \lambda BA$, $\lambda \notin \mathbb{R}$, has no *bounded* self-adjoint operators A and B verifying $AB \neq 0$. However, the relation $AB = \lambda BA$, with $|\lambda| = 1$, has representations by *unbounded* self-adjoint operators A and B (see [81] and [82]). Such unbounded operators are the "natural" representations of this relation.

The following result contains simpler proofs and certain generalizations of some of the results obtained in [10]:

Theorem 6.8.16 ([92]) *Let* $A, B \in B(H)$ *be such that* $AB = \lambda BA \neq 0$, $\lambda \in \mathbb{C}^*$. *Then*

1. *if* A *or* B *is self-adjoint, then* $\lambda \in \mathbb{R}$;
2. *if either* A *or* B *is self-adjoint and the other is normal, then* $\lambda \in \{-1, 1\}$; *and*
3. *if* A *and* B *are both normal, then* $|\lambda| = 1$.

There are different proofs of the previous theorem. Readers will see that some of these statements will be shown below (see, e.g., [10, 58, 92] for further details). So, we only provide a proof of the last statement:

Proof of (3) Chō et al. [14] Since A and B are normal, it may be shown that

$$\|AB\| = \|BA\|.$$

Since $\|AB\| \neq 0$, we immediately obtain $|\lambda| = 1$. $\qquad\square$

Theorem 6.8.17 *Let* A *and* B *be two bounded normal operators such that* $AB = \lambda BA \neq 0$, *where* $\lambda \in \mathbb{C}$. *Then* AB *(and also* BA*) is normal for any non-zero* λ.

Proof Since A and B are both normal, so are λA and λB. So by the Fuglede-Putnam theorem, we have

$$AB = \lambda BA \implies AB^* = \bar{\lambda} B^* A \text{ and } A^* B = \bar{\lambda} B A^*.$$

We have on the one hand

$$(AB)^* AB = B^* A^* AB = B^* A A^* B = \bar{\lambda} B^* A B A^* = |\lambda|^2 B^* B A A^*.$$

On the other hand we obtain

$$AB(AB)^* = ABB^* A^* = AB^* B A^* = \bar{\lambda} B^* A B A^* = |\lambda|^2 B^* B A A^*.$$

Thus AB is normal. $\qquad\square$

Remark Of course, thanks to Theorem 6.8.16, the condition $|\lambda| = 1$ was tacitly assumed in the previous theorem.

Remark By the above results, there cannot be two normal matrices A and B such that, e.g., $AB = 2BA$.

Corollary 6.8.18 *Let A and B be two bounded self-adjoint operators satisfying $AB = \lambda BA \neq 0$ where $\lambda \in \mathbb{C}$. If either $\sigma(A) \cap \sigma(-A) \subseteq \{0\}$ or $\sigma(B) \cap \sigma(-B) \subseteq \{0\}$, then $\lambda = 1$.*

Proof By Theorem 6.8.17, AB and BA are normal. By Theorem 6.8.2, AB and BA are self-adjoint. Hence

$$BA = (AB)^* = AB = \lambda BA,$$

yielding $\lambda = 1$. The proof is thus complete.   □

Corollary 6.8.19 *Let A and B be two bounded self-adjoint operators satisfying $AB = \lambda BA \neq 0$ where $\lambda \in \mathbb{C}$. Then $\lambda = 1$ if one of the following occurs:*

1. *A is positive;*
2. *$-A$ is positive;*
3. *B is positive;*
4. *$-B$ is positive.*

Corollary 6.8.20 *If both $A, B \in B(H)$ are self-adjoint, then*

$$AB = \lambda BA \Longrightarrow \lambda \in \{-1, 1\}.$$

Proof We have

$$A^2 B = \lambda ABA = \lambda^2 BA^2.$$

Since A is self-adjoint, A^2 is positive so that Corollary 6.8.19 gives $\lambda^2 = 1$ or $\lambda \in \{-1, 1\}$.   □

Now, we pass to the case where one operator is not necessarily bounded. The first attempt was [52], from which we choose to only give the following simple result.

Theorem 6.8.21 ([52]) *Let A be an unbounded operator and let $B \in B(H)$. Assume that $BA \subset \lambda AB \neq 0$ where $\lambda \in \mathbb{C}$. Then λ is real if A is self-adjoint.*

Proof ([52]) Since $BA \subset \lambda AB$ and since A is self-adjoint, the Fuglede-Putnam theorem yields $BA \subset \overline{\lambda} AB$. Now for $x \in D(A) = D(BA) \subset D(\lambda AB) = D(\overline{\lambda} AB)$, one has

$$\lambda ABx = \overline{\lambda} ABx.$$

Hence λ is real as $AB \neq 0$.   □

The next result generalizes Theorem 6.8.17.

Theorem 6.8.22 ([12]) *Let A and B be two normal operators where B is bounded. Assume that* $BA \subset \lambda AB \neq 0$ *where* $\lambda \in \mathbb{C}$. *Then*

$$AB \text{ is normal} \iff |\lambda| = 1.$$

Proof ([12]) First, since A is closed and B is bounded, AB is automatically closed. Since A is normal, so is λA. Hence the Fuglede-Putnam theorem gives

$$BA \subset \lambda AB \implies BA^* \subset \bar{\lambda} A^* B \text{ or } \lambda B^* A \subset AB^*.$$

Using the above "inclusions" we have on the one hand

$$(AB)^* AB \supset B^* A^* AB$$
$$= B^* AA^* B \text{ (since } A \text{ is normal)}$$
$$\supset \frac{1}{\lambda} B^* ABA^*$$
$$\supset \frac{1}{\bar{\lambda}\lambda} B^* BAA^*$$
$$= \frac{1}{|\lambda|^2} B^* BAA^*.$$

Since A and AB are closed, and B is bounded, all of $(AB)^* AB$, $A^* A$ and $B^* B$ are self-adjoint so that "adjointing" the previous inclusion yields

$$(AB)^* AB \subset \frac{1}{|\lambda|^2} AA^* B^* B.$$

As $|\lambda|$ is real, the conditions of Theorem 6.3.1 are all met, whereby

$$(AB)^* AB = \frac{1}{|\lambda|^2} AA^* B^* B.$$

On the other hand, we may write

$$AB(AB)^* \supset ABB^* A^*$$
$$= AB^* BA^* \text{ (because } B \text{ is normal)}$$
$$\supset \lambda B^* ABA^*$$
$$= B^* (\lambda AB) A^*$$
$$\supset B^* BAA^*.$$

As above, we obtain

$$AA^*B^*B \supset AB(AB)^*,$$

and by Theorem 6.3.1 once more, we end up with

$$AB(AB)^* = AA^*B^*B.$$

Accordingly, we clearly see that AB is normal iff $|\lambda| = 1$, as needed. □

Corollary 6.8.23 ([12]) *Let A and B be two normal operators where B is bounded. If $BA \subset \lambda AB \neq 0$ where $|\lambda| = 1$, then*

$$\overline{BA} = \lambda AB.$$

Proposition 6.8.24 ([12]) *Let A and B be two self-adjoint operators where B is bounded. Assume that $BA \subset \lambda AB \neq 0$ where $\lambda \in \mathbb{C}$. Then AB is normal for any λ.*

Proof *([12])* Since A and B are self-adjoint, $BA \subset \lambda AB$ implies the following three "inclusions"

$$BA \subset \overline{\lambda} AB, \ \lambda BA \subset AB \text{ and } \overline{\lambda} BA \subset AB.$$

Proceeding as in the proof of Theorem 6.8.22, we obtain

$$(AB)^*AB \supset |\lambda|^2 B^2 A^2 \text{ and } AB(AB)^* \supset |\lambda|^2 B^2 A^2.$$

Again, and as in the proof of Theorem 6.8.22, AB is normal. □

Corollary 6.8.25 *Let A and B be self-adjoint operators where $B \in B(H)$. Assume that $BA \subset \lambda AB \neq 0$ where $\lambda \in \mathbb{C}$. Then $\lambda = 1$ if B is positive.*

Proof *([12])* By Proposition 6.8.24, AB is normal. By Theorem 6.8.7, we get the self-adjointness of AB, i.e., $(AB)^* = AB$, and so

$$AB = (AB)^* \subset \frac{1}{\lambda} AB.$$

But $D(AB) = D(\alpha AB)$ for any $\alpha \neq 0$. Therefore,

$$AB = \frac{1}{\lambda} AB \text{ or merely } \lambda = 1.$$

□

Corollary 6.8.26 *Let A and B be self-adjoint operators where B is bounded. Assume that $BA \subset \lambda AB \neq 0$ where $\lambda \in \mathbb{C}$. Then $\lambda \in \{-1, 1\}$.*

Proof We may write

$$B^2 A \subset \lambda BAB \subset \lambda^2 AB^2.$$

Since B is self-adjoint, B^2 is positive so that Corollary 6.8.25 applies and gives $\lambda^2 = 1$ or $\lambda \in \{-1, 1\}$. $\qquad\square$

What about the case where the unbounded self-adjoint operator A is positive? This was an open problem from [12], and a positive answer appeared thereafter in [30]:

Proposition 6.8.27 *Let A and B be self-adjoint operators where $B \in B(H)$. Assume that $BA \subset \lambda AB \neq 0$ where $\lambda \in \mathbb{C}$. Then $\lambda = 1$ if A is positive.*

Proof By Proposition 6.8.24, we already know that AB is normal. By Theorem 6.8.10, AB is then self-adjoint. Now,

$$BA \subset \lambda AB \implies \frac{1}{\lambda} BA \subset AB.$$

Hence

$$AB = (AB)^* \subset \frac{1}{\lambda} AB.$$

Therefore,

$$AB = \frac{1}{\lambda} AB \text{ or simply } \lambda = 1.$$

$\qquad\square$

Now, we deal with a similar treatment of the problem when both A and B are normal and unbounded. As readers expect, this will be carried out under relatively strong conditions. First, and as a consequence of Theorem 5.1.10, we have:

Corollary 6.8.28 *If A and B are two unbounded normal and invertible operators such that $AB = \lambda BA$, then*

$$A^* B = \overline{\lambda} B A^* \text{ and } A B^* = \overline{\lambda} B^* A.$$

The following result may therefore be shown using the above methods.

Theorem 6.8.29 ([12]) *Let A, B two unbounded invertible operators such that $AB = \lambda BA \neq 0$, $\lambda \in \mathbb{C}$. Then*

1. If A or B is self-adjoint, then $\lambda \in \mathbb{R}$.
2. If either A or B is self-adjoint and the other is normal, then $\lambda \in \{-1, 1\}$.
3. If both A and B are normal, then $|\lambda| = 1$.

The remaining part of this section is devoted to the sum of two λ-commuting operators. However, there is not much one can do apart from one or two results.

We already know that the sum of two bounded commuting normal operators is normal. It is also known that the sum of two strongly anti-commuting unbounded self-adjoint operators is self-adjoint (see [88]). This result, however, cannot naively be generalized to the case of two anticommuting normal 2×2 matrices, let alone the case of two unbounded normal operators. Indeed,

Example 6.8.30 ([57]) Let

$$A = \begin{pmatrix} 2 & 0 \\ 0 & -2 \end{pmatrix} \text{ and } B = \begin{pmatrix} 0 & 1 \\ -1 & 0 \end{pmatrix}.$$

Then A is self-adjoint and B is unitary. They anticommute because

$$AB = -BA = \begin{pmatrix} 0 & 2 \\ 2 & 0 \end{pmatrix},$$

but as one can easily check

$$A + B = \begin{pmatrix} 2 & 1 \\ -1 & -2 \end{pmatrix}$$

is not normal.

Nevertheless, we have the following result, which is perhaps unexciting.

Theorem 6.8.31 ([57]) *Let A be an unbounded normal operator and let B be bounded and self-adjoint. If $BA^* \subset -AB$, then $A + B$ is normal.*

6.9   Algebras of Normal Operators

Theorem 6.9.1 ([73]) *Let $N \subset B(H)$ be constituted of normal operators, where H is a complex Hilbert space. If N is a vector space, then each pair in N is commutative.*

Proof ([73]) Let $A, B \in N$. Since N is a vector space, $A + B$ and $A + iB$ ($i^2 = -1$) are both in N, i.e., $A + B$ and $A + iB$ are both normal. If we set

$$C = (A + B)(A + B)^* - (A + B)^*(A + B)$$

and

$$D = (A + iB)(A + iB)^* - (A + iB)^*(A + iB),$$

then $C = D = 0$. Hence $C + iD = 0$. On expansion

$$AB^* + BA^* - A^*B - B^*A + i(-iAB^* + iBA^* - iA^*B + iB^*A) = 0.$$

After simplification, we get

$$2(AB^* - B^*A) = 0 \text{ or } AB^* = B^*A,$$

whereby $AB = BA$ by the Fuglede theorem. This completes the proof.   □

Remark Readers might also be interested in consulting [45].

6.10   A Result by F. Kittaneh

Theorem 6.10.1 ([41], cf. [44]) *Let $T = A + iB \in B(H)$ be hyponormal, where A and B are self-adjoint. If AB is hyponormal, then T is normal.*

Proof *([41])* Since T is hyponormal, we have

$$T^*T - TT^* = 2i(AB - BA) = 2i[AB - (AB)^*] \geq 0.$$

Because A and B are self-adjoint, we may write

$$(AB)A = A(BA) = A(AB)^*.$$

Since AB is hyponormal, Theorem 2.3.2 gives

$$(AB)^*A = A(AB)^{**} = A(AB).$$

Hence

$$[AB + (AB)^*]A = A[AB + (AB)^*].$$

Also,

$$[AB - (AB)^*]A = A[(AB)^* - AB].$$

By setting $C = i[AB - (AB)^*]$, we see that

$$CA = -AC \text{ and so } C^2A = AC^2.$$

Since $C \geq 0$, we obtain $CA = AC$, or that $(AB)^* - AB$ commutes with A. Therefore, A commutes with AB for$(AB)^* + AB$ commutes with A.

Therefore,

$$A(AB - BA) = (AB - BA)A = 0.$$

A well known result tells us that $\sigma(AB - BA) = \{0\}$. In fine, as $AB - BA$ is normal, it follows that

$$AB - BA = 0,$$

establishing the normality of T, as needed.   □

6.11 Roots of Normal Operators

Recall that $B \in B(H)$ is called a square root of some $A \in B(H)$ when $B^2 = A$. It is well-known that an operator may have no square root at all, as it may have infinitely many of them, and also anything between these two extreme cases (see, e.g., [58] or [62]). Observe that a normal operator may have a non-normal square root. For example, the normal matrix $\begin{pmatrix} 0 & 0 \\ 0 & 0 \end{pmatrix}$ defined on a bi-dimensional space has, e.g., $\begin{pmatrix} 0 & 1 \\ 0 & 0 \end{pmatrix}$, which is not normal, as one of its square roots.

However, a normal operator always possesses square roots (regardless of the dimension of the Hilbert space and regardless of the boundedness of the operator). The obvious way to see that is to invoke the spectral theorem. Indeed, if N is normal, then it may be represented as $\int \lambda dE$. So, $B = \int \sqrt{\lambda} dE$, where $\sqrt{\lambda}$ is either of the two square roots of the complex λ, is a square root of N. Another less known way based upon elementary operator theory is given next:

Theorem 6.11.1 ([61]) *Let $N = A + iB \in B(H)$ be normal such that say $B \geq 0$. Then*

$$T = \left(\frac{|N| + A}{2} \right)^{\frac{1}{2}} + i \left(\frac{|N| - A}{2} \right)^{\frac{1}{2}}$$

is a normal square root of N.

Remark A similar formula exists for $B \leq 0$.

That said, one can now ask under what conditions a square root of a normal operator is itself normal? C.R. Putnam provided the following answer (see [74] for deeper results):

Theorem 6.11.2 ([71]) *Let N be a fixed bounded normal operator, and let A be an arbitrary solution of $A^2 = N$. Suppose that there is a line L in the complex plane passing through the origin and lying entirely on one side of (and possible lying all, or partly, in) the set $\overline{W(A)}$. Then A must be normal.*

Proof ([71]) First, choose an angle α for which the operator $B = e^{i\alpha}A$ satisfies $B + B^* \geq 0$. Let $B = H + iJ$ be the Cartesian decomposition of B, and so $H = (B + B^*)/2$ and $J = (B - B^*)/2i$. It is easy to see that

$$B^2 = H^2 - J^2 + i(HJ + JH).$$

Since A^2 is normal, $e^{2i\alpha}A^2$ or B^2 are normal too. Since $B^2B = BB^2$, the Fuglede theorem gives $(B^2)^*B = B(B^2)^*$ or $B^2B^* = B^*B^2$. Hence $HB^2 = B^2H$ and $JB^2 = B^2J$. The above displayed equation multiplied on the left by H becomes

$$HB^2 = H^3 - HJ^2 + i(H^2J + HJH).$$

Also

$$B^2H = H^3 - J^2H + i(HJH + JH^2).$$

A subtraction of the two equations now implies $R + iS = 0$, where $R = J^2H - HJ^2$ and $S = H^2J - JH^2$. Since R and S are not self-adjoint, we cannot directly obtain $R = S = 0$. However, and as $R^* = -R$ and $S^* = -S$, we then get

$$0 = R^* - iS^* = -R + iS \Longrightarrow R - iS = 0.$$

Therefore, $S = 0$ or $H^2J = JH^2$. Since $H \geq 0$, it ensues that $HJ = JH$. Accordingly, B is normal, thereby A is normal too, as needed.   □

Remark The previous result tells in particular that if $A^2 = N$, N is normal and $\text{Re } A \geq 0$, then A is normal. Somehow similar results appeared in [61] (cf. [25]).

A related result about normal square roots reads:

Theorem 6.11.3 ([42], cf. Stampfli [84]) *Let $B, N \in B(H)$ be commuting such that B is normal and $N^2 = 0$. Set $A = B + N$. If A is invertible and A^2 is normal, then A is normal.*

Proof ([42]) Clearly

$$A^2 = (B + N)^2 = B^2 + NB + BN + N^2 = B^2 + 2BN$$

because N commutes with B and $N^2 = 0$, and so $A^2 - B^2 = 2BN$.

Second, we may write

$$AB = BA \implies AB^2 = BAB = B^2A$$
$$\implies A^2B^2 = AB^2A = B^2A^2$$
$$\implies A^2(-B^2) = (-B^2)A^2.$$

By Theorem 6.6.1, we obtain the normality of $A^2 - B^2$, whereby BN is normal.

Now, by Exercise 7.3.29 in [58], we have $\sigma(A) = \sigma(B)$ for N is nilpotent. In particular, B is invertible because A is. Besides,

$$NB = BN \implies N = BNB^{-1} (= B^{-1}BN).$$

But B^{-1} is normal, and since BN is normal (and commutes with the normal B^{-1}), we infer by Theorem 6.5.2 that N is normal. Therefore, we obtain $N = 0$ as $N^2 = 0$. Thus $A = B$, i.e., A is normal.   □

As a consequence of M.R. Embry's theorem, we have:

Proposition 6.11.4 ([23]) *Let $A \in B(H)$ be such that $0 \notin W(A)$. If A^2 is normal, then A too is normal.*

In the end, the following related result seems noteworthy (we do include a proof for it uses an interesting argument):

Theorem 6.11.5 ([17]) *Let A be a boundedly invertible non-necessarily bounded operator. If A^p and A^q are normal, where p and q are two co-prime numbers, then A is normal.*

Proof *([17])* We have by Bézout's theorem in arithmetic that $ap + bq = 1$ for some integers a and b. Necessarily, one of ap and bq has to be negative, and WLOG assume it is bq. Since A is boundedly invertible, we have $A^{-1}A \subset I$ and $AA^{-1} = I$. Hence

$$A^{bq}A^{ap} \subset A = A^{ap}A^{bq}.$$

Since A is boundedly invertible, $A^{bq} \in B(H)$. Since A^p and A^q are also normal, A^{ap} and A^{bq} remain normal. In other words, the bounded and normal A^{bq} commutes with the normal A^{ap}. Hence, by say Theorem 6.5.3,

$$A^{ap}A^{bq} = A$$

is normal, as needed.   □

An akin result holds for unbounded self-adjoint operators as well as unbounded self-adjoint positive operators. A proof may be obtained using Theorem 6.5.1, and so we leave details of the proof to readers.

Theorem 6.11.6 ([17]) *Let A be a boundedly invertible non-necessarily bounded operator. If A^p and A^q are self-adjoint (resp. self-adjoint and positive), where p and q are two co-prime numbers, then A is self-adjoint (resp. self-adjoint and positive).*

Remark The condition "p and q being co-prime numbers" cannot just be dropped. For example, consider an invertible non-normal square root A of the identity matrix, and then $A^2 = A^4 = I$. An explicit example would be the non-normal
$$A = \begin{pmatrix} 2 & 1 \\ -3 & -2 \end{pmatrix}.$$

Corollary 6.11.7 *Let A be a boundedly invertible but non-necessarily bounded operator. If A^p and A^q are normal, $TA^p \subset A^p T$, $TA^q \subset A^q T$, and p and q are two co-prime numbers, then $TA \subset AT$, so $TA^* \subset A^*T$.*

Proof The conclusion $TA \subset AT$ may be reached without the normality of A^p and A^q (as carried out in say Proposition 3.22 in [17]). Then, apply the Fuglede theorem. □

By Berberian's trick, we infer the following result, whose proof is left as an exercise for interested readers.

Corollary 6.11.8 *Let A and B be two boundedly invertible but non-necessarily bounded operators. If $TA^p \subset B^p T$, $TA^q \subset B^q T$, and p and q are two co-prime numbers, then $TA \subset BT$. Hence $TA^* \subset B^*T$ if A and B are also normal.*

Remark Readers are very encouraged to consult [2, 22, 24], and [43] for some related results. For the case of unbounded operators, a relevant article is [11]. The latter contains an interesting result that is unknown, unfortunately, to many specialists. For example, it will be used

Remark The finding in [11] may be utilized to show the following result (which also uses the Fuglede-Putnam theorem): *Let A be an unbounded, densely defined, and symmetric operator; and let $B \in B(H)$ be positive such that BA^2 is normal. Then $BA^2 \subset A^2 B$. If it is further assumed that $\sigma(A) \subset (0, \infty)$, then $BA \subset AB$.*
Details will appear elsewhere.

6.12 Radjavi-Rosenthal's Theorem

Theorem 6.12.1 ([74]) *Let $A, B, X \in B(H)$ be such that A and B^* are subnormal and X is positive and one-to-one (hence X has dense range). If $XB = AX$, then A and B are normal and $A = B$.*

Proof *([74])* By the subnormality of A and B^*, we know that there are normal operators N and M such that

$$N = \begin{pmatrix} A & R \\ 0 & C \end{pmatrix} \text{ and } M = \begin{pmatrix} B & 0 \\ S & D \end{pmatrix}$$

on $K := H \oplus H_1$, where H_1 is a Hilbert space. Define now on K the operator matrix $\widetilde{X} = \begin{pmatrix} X & 0 \\ 0 & 0 \end{pmatrix}$. The hypothesis $XB = AX$ yields $N\widetilde{X} = \widetilde{X}M$. The Fuglede-Putnam theorem then gives $N^*\widetilde{X} = \widetilde{X}M^*$. By taking the adjoint of both sides, and remembering that $\widetilde{X}$ is positive, in particular self-adjoint, we get $M\widetilde{X} = \widetilde{X}N$. Therefore,

$$XA = BX, \ XR = 0 \text{ and } SX = 0.$$

Since X is injective and has dense range, we obtain $R = 0$ and $S = 0$. Thus, we have forced A and B to be normal. The equations $XB = AX$ and $XA = BX$ imply that $X^2A = AX^2$. Since $X \geq 0$, it follows that $XA = AX$ or merely that $AX = BX$, thereby $A = B$, as needed.   □

The following consequence appeared in [74] with a slightly different assumption:

Corollary 6.12.2 *If $A, B, X \in B(H)$, where A and B^* are subnormal and X is invertible, and if $XB = AX$, then A and B are normal and unitarily equivalent.*

Proof Since X is invertible, we know that $X = U|X|$ for a certain unitary operator U. Since $XB = AX$, it follows that $|X|B = U^*AU|X|$. Since A is subnormal, so is U^*AU. So the foregoing result tells us that U^*AU and B are, in fact, normal and that $U^*AU = B$, thus proving that A is unitarily equivalent to B, as needed.   □

Remark See [85] and [86] for some related results.

6.13   Certain Maximality Results

First, what do we mean by maximality results here? Well, these are theoretical results containing conditions that force an "inclusion" between two operators to become full equality.

We already know that self-adjoint operators are maximally symmetric, and normal operators are maximally normal. The latter properties have been used in the present manuscript on several occasions.

These two results are simple consequences of the following more general result:

Proposition 6.13.1 *Let S and T be two operators with domains $D(S)$ and $D(T)$ respectively. If S is densely defined and $D(S^*) \subset D(S)$, then*

$$S \subset T \implies S = T$$

whenever $D(T) \subset D(T^*)$.

Proof Since $S \subset T$, $T^* \subset S^*$. The hypotheses $D(S^*) \subset D(S)$ and $D(T) \subset D(T^*)$ then yield $S = T$.   □

Let us agree to call the maximality result of the previous proposition the maximality of operators of the first kind. There are other types of maximality. For instance, we may well have a product of two operators on either side of the inclusion (we shall call that maximality of operators of the second kind). An illustration of that being Theorem 6.3.1 met above (see [65] for some related results).

There is yet another akin situation that might be useful as well. This is referred to in [57], and here too, as a double maximality of operators for it involves a system of two inclusions. More results may be consulted in [46, 57, 87, 89].

Before focusing our attention on Fuglede-Putnam-related proofs, it seems that the following double maximality result could be of some interest to readers. Notice that most of the remaining of this section is borrowed from [46].

Proposition 6.13.2 *Let S be a densely defined operator such that $S \subset T$ and $S \subset T^*$. If $D(T) = D(T^*)$, then T is self-adjoint.*

Proof For all $x \in D(T) = D(T^*)$ and for all $y \in D(S) \subset D(T) = D(T^*)$ we may write

$$\langle Tx, y \rangle = \langle x, T^*y \rangle = \langle x, Sy \rangle = \langle x, Ty \rangle = \langle T^*x, y \rangle.$$

Since $D(S)$ is dense, it follows that $Tx = T^*x$ for all $x \in D(T) = D(T^*)$, that is, T is self-adjoint.   □

Corollary 6.13.3 *Let S be a densely defined operator such that $S \subset T$ and $S \subset T^*$. If T is normal, then it is self-adjoint.*

Let us now turn to some generalizations of maximality results of the second kind where one operator is normal. As noted in [19], there cannot be a version of Theorem 6.3.1 solely involving normal operators. Also, there is no hope for a maximality result of the third kind, i.e., a version of the product of three operators, to hold even for self-adjoint operators (see [46] for a counterexample).

Now, we list some of the results involving normal operators. The use of the Fuglede-Putnam theorem seems inevitable. The reason is that in some cases, one of the operators (or its adjoint) is symmetric, and there is no spectral theorem (which has been the alternative to the Fuglede theorem in several cases) for such a class. In the end, note that the remaining material of this section is borrowed from [46].

Theorem 6.13.4 *Let A, B and T be linear operators such that $B \in B(H)$. If T^* is symmetric, B is self-adjoint and A is normal, then*

$$T \subset AB \Longrightarrow \overline{T} = AB.$$

In particular, if we further assume that T is closed, then we obtain $T = AB$.

Remark It would be interesting to investigate whether one can obtain a version of the preceding theorem involving two normal operators.

Proof Since AB is closed, it ensues that

$$T \subset AB \Longrightarrow \overline{T} \subset AB.$$

Hence

$$T \subset AB \Longrightarrow BA^* \subset (AB)^* \subset T^* \subset T^{**} = \overline{T} \subset AB.$$

The Fuglede-Putnam theorem then gives $BA \subset A^*B$. So

$$B^2 A^* A \subset BABA \subset BAA^*B = B^*A^*AB \subset (AB)^*AB.$$

Similarly,

$$B^2 AA^* \subset AB(AB)^* \text{ or } B^2 A^*A \subset AB(AB)^*$$

thanks to the normality of A. Since $B^2 \in B(H)$, the self-adjointness of all of A^*A, $AB(AB)^*$, $(AB)^*AB$ and B^2 entail

$$AB(AB)^* \subset A^*AB^2 \text{ and } (AB)^*AB \subset A^*AB^2.$$

Theorem 6.3.1 then gives

$$(AB)^*AB = AB(AB)^*,$$

i.e., AB is normal. Hence $(AB)^*$ too is normal. Since normal operators are maximally symmetric (by Proposition 6.13.1), we get

$$(AB)^* \subset T^* \Longrightarrow (AB)^* = T^* \Longrightarrow AB = \overline{AB} = (AB)^{**} = T^{**} = \overline{T},$$

as needed.   □

Remark In general,

$$BA \subset T \not\Longrightarrow BA = T$$

even when A, B and T are all self-adjoint. Indeed, just consider a boundedly invertible self-adjoint A with a domain $D(A) \subsetneq H$ such that $A^{-1} = B \in B(H)$ and $T = I_H$ (the identity operator on the entire space H). Then

$$BA = A^{-1}A = I_{D(A)} \subsetneq I_H = T,$$

where $I_{D(A)}$ is the identity operator on $D(A)$.

Corollary 6.13.5 *Let A, B and T be linear operators such that $B \in B(H)$. If T is symmetric, B is self-adjoint and A is normal, then*

$$BA \subset T \Longrightarrow \overline{T} = \overline{BA}.$$

Proof As above, we get

$$\overline{BA}(BA)^* = (BA)^*\overline{BA} \ (= A^*AB^2).$$

Since normal operators are maximally symmetric, we obtain

$$BA \subset T \Longrightarrow \overline{BA} \subset \overline{T} \Longrightarrow \overline{T} = \overline{BA},$$

as required. □

6.14 Counterexamples

First, we provide an example of two similar hyponormal (resp. subnormal) operators that are not unitarily equivalent.

Example 6.14.1 Most of the material here is borrowed from [31] with some adjustments. It is perhaps better to give the operators in their matrix representations. Let

$$R = \begin{bmatrix} 0 & 0 & 0 \cdots & \cdots \\ 1/\sqrt{2} & 0 & 0 & 0 \\ 0 & 1 & 0 & 0 & \ddots \\ & \ddots & 1 & 0 & 0 & \ddots \\ & & & 1 & \ddots & \ddots \\ \vdots & & & & \ddots & \ddots \end{bmatrix} \quad \text{and } S = \begin{bmatrix} 0 & 0 & 0 \cdots & \cdots \\ 1 & 0 & 0 & 0 \\ 0 & 1 & 0 & 0 & \ddots \\ & \ddots & 1 & 0 & 0 & \ddots \\ & & & 1 & \ddots & \ddots \\ \vdots & & & & \ddots & \ddots \end{bmatrix},$$

i.e., S is the usual shift and R is a (unilateral) weighted shift, both defined on ℓ^2. Then both S and R are hyponormal (in fact, subnormal).

The similarity of S and R is obtained via the invertible operator

$$T = \begin{bmatrix} 1 & 0 & & & \cdots & \\ 0 & \sqrt{2} & 0 & & \cdots & \\ 0 & 0 & \sqrt{2} & \ddots & & \ddots \\ & \ddots & \ddots & \sqrt{2} & & \ddots \\ & & & & \ddots & \ddots \\ \vdots & & & & \ddots & \ddots \end{bmatrix},$$

that is, $TR = ST$.

That are S and R not unitarily equivalent is left to readers.

Remark ([62]) We have already seen that similar subnormal operators do not have to be unitarily equivalent. The corresponding question for quasinormal had resisted solutions for almost 15 years since the paper [40]. It was then solved in the negative in [13] (one of the authors of the previous reference also announced the counterexample in [90]).

Next, we supply a counterexample to the reverse implication in Corollary 6.2.3:

Example 6.14.2 ([4]) Let A and B be acting on the standard basis (e_n) of $\ell^2(\mathbb{N})$ by:

$$Ae_n = \alpha_n e_n \text{ and } Be_n = e_{n+1}, \ \forall n \geq 1$$

respectively. Assume further that α_n is bounded, *real-valued* and *positive*, for all n. Hence A is self-adjoint (and positive). Then

$$ABe_n = \alpha_{n+1} e_{n+1} \text{ and } BAe_n = \alpha_n e_{n+1}, \ \forall n \geq 1,$$

meaning that both AB and BA are weighted shifts with weights $\{\alpha_n\}_{n=2}^\infty$ and $\{\alpha_n\}_{n=1}^\infty$ respectively.

Now recall that (see e.g., [31]): a weighted shift with weight $\{\alpha_n\}_{n=1}^\infty$ is hyponormal if and only if $\{|\alpha_n|\}_{n=1}^\infty$ is monotonically increasing.

Hence if $\{\alpha_n\}_{n=1}^\infty$ is increasing, then AB and BA are both hyponormal. Moreover, $AB = BA$ (equivalently, $A^2B = BA^2$) if and only if the sequence $\{\alpha_n\}_{n=1}^\infty$ is constant. Taking any nonconstant monotonically increasing sequence $\{\alpha_n\}_{n=1}^\infty$ does the job.

Next, we supply two unbounded self-adjoint operators A and B such that A is positive, AB is normal but non-self-adjoint (cf. Theorem 6.8.11):

Example 6.14.3 ([50, 62]) Let $B = -i\frac{d}{dx}$. Then B is self-adjoint on $H^1(\mathbb{R})$. Take the self-adjoint positive $Af(x) = (1 + |x|)f(x)$ with the following domain

$$D(A) = \{f \in L^2(\mathbb{R}) : (1 + |x|)f \in L^2(\mathbb{R})\}.$$

First, we check that $M := AB$ is normal, i.e., M is closed and $MM^* = M^*M$. That M is closed follows from the (bounded) invertibility of A and the closedness of B.

For f in $D(M) = \{f \in L^2(\mathbb{R}) : (1 + |x|)f' \in L^2(\mathbb{R})\}$, one has

$$Mf(x) = -i(1 + |x|)f'(x).$$

Then

$$M^*f(x) = \mp if(x) - i(1 + |x|)f'(x).$$

Clearly

$$MM^*f(x) = M^*Mf(x) = -(1 + |x|)(\mp 2f'(x) - (1 + |x|)f''(x)).$$

The previous equations combined with

$$D(MM^*) = D(M^*M) = \{f \in L^2(\mathbb{R}) : (1 + |x|)f', (1 + |x|)f'' \in L^2(\mathbb{R})\}$$

say that AB is a normal operator. Observe in the end that M is not self-adjoint.

The last example to be given is a pair of an unbounded self-adjoint operator A and a positive unbounded self-adjoint operator B such that $\overline{AB}$ is normal, but it is not self-adjoint, i.e., AB is not essentially self-adjoint. The counterexample is given without details (these may be consulted in either [48, 49] or [62]):

Example 6.14.4 Consider the operators A and B defined as:

$$A = -i\frac{d}{dx} : D(A) \to L^2(\mathbb{R}), \ B = |x| : D(B) \to L^2(\mathbb{R})$$

where $D(B) = \{f \in L^2(\mathbb{R}) : xf \in L^2(\mathbb{R})\}$ and $D(A) = H^1(\mathbb{R})$. Then $N := AB$ is defined by

$$Nf(x) = -i(|x|f(x))' = -i|x|f'(x) - i \ \text{sign}(x)f(x)$$

for f in $D(N) = \{f \in L^2(\mathbb{R}) : |x|f, -i(|x|f)' \in L^2(\mathbb{R})\}$. Then AB has the required properties.

References

1. E. Albrecht, P.G. Spain, When products of self-adjoints are normal. Proc. Am. Math. Soc. **128**(8), 2509–2511 (2000)
2. A.H. Al-Moajil, The commutants of relatively prime powers in Banach algebras. Proc. Am. Math. Soc. **57**(2), 243–249 (1976)
3. T. Ando, Operators with a norm condition. Acta Sci. Math. (Szeged) **33**, 169–178 (1972)
4. A. Benali, M.H. Mortad, Generalizations of Kaplansky's theorem involving unbounded linear operators. Bull. Pol. Acad. Sci. Math. **62**(2), 181–186 (2014)
5. S.K. Berberian, Note on a theorem of Fuglede and Putnam. Proc. Am. Math. Soc. **10**, 175–182 (1959)
6. S.J. Bernau, The square root of a positive self-adjoint operator. J. Aust. Math. Soc. **8**, 17–36 (1968)
7. M.S. Birman, M.Z. Solomjak, Spectral theory of selfadjoint operators in Hilbert space. Translated from the 1980 Russian original by S. Khrushchëv, V. Peller, *Mathematics and its Applications (Soviet Series)* (D. Reidel Publishing, Dordrecht, 1987)
8. E. Bishop, Spectral theory for operators on a Banach space. Trans. Am. Math. Soc. **86**, 414–445 (1957)
9. I. Boucif, S. Dehimi, M.H. Mortad, On the absolute value of unbounded operators. J. Oper. Theory **82**(2), 285–306 (2019)
10. J.A. Brooke, P. Busch, D.B. Pearson, Commutativity up to a factor of bounded operators in complex Hilbert space. R. Soc. Lond. Proc. Ser. A Math. Phys. Eng. Sci. **458**(2017), 109–118 (2002)
11. S.R. Caradus, Commutation properties of operator polynomials. J. Aust. Math. Soc. **12**, 98–100 (1971)
12. C. Chellali, M.H. Mortad, Commutativity up to a factor for bounded and unbounded operators. J. Math. Anal. Appl. **419**(1), 114–122 (2014)
13. K.Y. Chen, D.A. Herrero, P.Y. Wu, Similarity and quasisimilarity of quasinormal operators. J. Oper. Theory **27**(2), 385–412 (1992)
14. M. Chō, J.I. Lee, T. Yamazaki, On the operator equation $AB = zBA$. Sci. Math. Jpn., **69**(2), 257–263 (2009)
15. S. Dehimi, M.H. Mortad, Right (or left) invertibility of bounded and unbounded operators and applications to the spectrum of products. Complex Anal. Oper. Theory **12**(3), 589–597 (2018)
16. S. Dehimi, M.H. Mortad, Generalizations of Reid Inequality. Math. Slovaca **68**(6), 1439–1446 (2018)
17. S. Dehimi, M.H. Mortad, Unbounded operators having self-adjoint, subnormal or hyponormal powers. Math. Nachr. (to appear)
18. S. Dehimi, M.H. Mortad, A. Bachir, On the commutativity of closed symmetric operators (submitted). arXiv:2203.07266
19. A. Devinatz, A.E. Nussbaum, On the permutability of normal operators. Ann. Math. **65**(2), 144–152 (1957)
20. A. Devinatz, A.E. Nussbaum, J. von Neumann, On the permutability of self-adjoint operators. Ann. Math. **62**(2), 199–203 (1955)
21. R.G. Douglas, On the operator equation $S^*XT = X$ and related topics. Acta Sci. Math. (Szeged) **30**, 19–32 (1969)
22. M.R. Embry, nth roots of operators. Proc. Am. Math. Soc. **19**, 63–68 (1968)
23. M.R. Embry, Similirates involving normal operators on Hilbert space. Pac. J. Math. **35**(2), 331–336 (1970)
24. M. Finkelstein, A. Lebow, A note on "nth roots of operators". Proc. Am. Math. Soc. **21**, 250 (1969)
25. N. Frid, M.H. Mortad, S. Dehimi, When nilpotence implies the zeroness of linear operators. Khayyam J. Math. **8**(2), 163–173 (2022)

26. B. Fuglede, A commutativity theorem for normal operators. Proc. Nati. Acad. Sci. **36**, 35–40 (1950)
27. S.R. Garcia, R.A. Horn, in *A Second Course in Linear Algebra*. Cambridge Mathematical Textbooks (Cambridge University Press, Cambridge, 2017)
28. A. Gheondea, When are the products of normal operators normal? Bull. Math. Soc. Sci. Math. Roum. (N.S.) **52(100)**(2), 129–150 (2009)
29. K. Gustafson, M.H. Mortad, Unbounded products of operators and connections to Dirac-type operators. Bull. Sci. Math. **138**(5), 626–642 (2014)
30. K. Gustafson, M.H. Mortad, Conditions implying commutativity of unbounded self-adjoint operators and related topics. J. Oper. Theory **76**(1), 159–169 (2016)
31. P.R. Halmos, *A Hilbert Space Problem Book*, 2nd edn. (Springer, Berlin, 1982)
32. R. Harte, The triangle inequality in C^*-algebras. Filomat **20**(2), 51–53 (2006)
33. M. Hladnik, M. Omladič, Spectrum of the product of operators. Proc. Am. Math. Soc. **102**(2), 300–302 (1988)
34. Z.J. Jabłoński, I.B. Jung, J. Stochel, Unbounded quasinormal operators revisited. Integr. Equ. Oper. Theory **79**(1), 135–149 (2014)
35. J. Janas, On unbounded hyponormal operators III. Stud. Math. **112**(1), 75–82 (1994)
36. C.-C. Jiang, On products of two Hermitian operators. Linear Algebra Appl. **430**(1), 553–557 (2009)
37. I.B. Jung, M.H. Mortad, J. Stochel, On normal products of selfadjoint operators. Kyungpook Math. J. **57**, 457–471 (2017)
38. I. Kaplansky, Products of normal operators. Duke Math. J. **20**(2), 257–260 (1953)
39. T. Kato, *Perturbation Theory for Linear Operators*, 2nd edn. (Springer, Berlin, 1980)
40. S. Khasbardar, N. Thakare, Some counterexamples for quasinormal operators and related results. Indian J. Pure Appl. Math. **9**(12), 1263–1270 (1978)
41. F. Kittaneh, On generalized Fuglede-Putnam theorems of Hilbert-Schmidt type. Proc. Am. Math. Soc. **88**(2), 293–298 (1983)
42. F. Kittaneh, On normality of operators. Rev. Roum. Math. Pures Appl. **29**(8), 703–705 (1984)
43. F. Kittaneh, On the commutants modulo C_p of A^2 and A^3. J. Aust. Math. Soc. Ser. A **41**(1), 47–50 (1986)
44. C.S. Lin, M. Radjabalipour, On intertwining and factorization by self-adjoint operators. Canad. Math. Bull. **21/1**, 47–51 (1978)
45. G. Maxwell, Algebras of normal matrices. Pac. J. Math. **43**, 421–428 (1972)
46. M. Meziane, M.H. Mortad, Maximality of linear operators. Rend. Circ. Mat. Palermo, Ser II. **68**(3), 441–451 (2019)
47. M. Möller, On the essential spectrum of a class of operators in Hilbert space. Math. Nachr. **194**, 185–196 (1998)
48. M.H. Mortad, An application of the Putnam-Fuglede theorem to normal products of self-adjoint operators. Proc. Am. Math. Soc. **131**(10), 3135–3141 (2003)
49. M.H. Mortad, *Normal Products of Self-Adjoint Operators and Self-Adjointness of the Perturbed Wave Operator on $L^2(\mathbb{R}^n)$*. Ph.D. Thesis. The University of Edinburgh. ProQuest LLC, Ann Arbor, 2003
50. M.H. Mortad, On some product of two unbounded self-adjoint operators. Integr. Equ. Oper. Theory **64**(3), 399–408 (2009)
51. M.H. Mortad, Yet more versions of the Fuglede-Putnam theorem. Glasgow Math. J. **51**(3), 473–480 (2009)
52. M.H. Mortad, Commutativity up to a factor: more results and the unbounded case. Z. Anal. Anwendungen: J. Anal. Appl. **29**(3), 303–307 (2010)
53. M.H. Mortad, Similarities involving unbounded normal operators. Tsukuba J. Math. **34**(1), 129–136 (2010)
54. M.H. Mortad, Products and sums of bounded and unbounded normal operators: Fuglede-Putnam versus Embry. Rev. Roum. Math. Pures Appl. **56**(3), 195–205 (2011)
55. M.H. Mortad, On the normality of the sum of two normal operators. Complex Anal. Oper. Theory **6**(1), 105–112 (2012)

56. M.H. Mortad, Commutativity of unbounded normal and self-adjoint operators and applications. Oper. Matrices **8**(2), 563–571 (2014)
57. M.H. Mortad, A criterion for the normality of unbounded operators and applications to self-adjointness. Rend. Circ. Mat. Palermo (2) **64**(1), 149–156 (2015)
58. M.H. Mortad, *An Operator Theory Problem Book* (World Scientific Publishing, Singapore, 2018)
59. M.H. Mortad, On the absolute value of the product and the sum of linear operators. Rend. Circ. Mat. Palermo II **68**(2), 247–257 (2019)
60. M.H. Mortad, On the invertibility of the sum of operators. Anal. Math. **46**(1), 133–145 (2020)
61. M.H. Mortad, On the existence of normal square and nth roots of operators. J. Anal. **28**(3), 695–703 (2020)
62. M.H. Mortad, *Counterexamples in Operator Theory* (Birkhäuser/Springer, Cham, 2022)
63. M.H. Mortad, Counterexamples related to unbounded paranormal operators. Examples Counterexamples (to appear). https://doi.org/10.1016/j.exco.2021.100017
64. E. Nelson, Analytic vectors. Ann. Math. **70**(2), 572–615 (1959)
65. A.E. Nussbaum, A commutativity theorem for unbounded operators in Hilbert space. Trans. Am. Math. Soc. **140**, 485–491 (1969)
66. S. Ôta, K. Schmüdgen, On some classes of unbounded operators. Integr. Equ. Oper. Theory **12**(2), 211–226 (1989)
67. A.B. Patel, P.B. Ramanujan, On sum and product of normal operators. Indian J. Pure Appl. Math. **12**(10), 1213–1218 (1981)
68. G.K. Pedersen, Analysis now, in *Graduate Texts in Mathematics*, vol. 118 (Springer-Verlag, New York, 1989)
69. A. Planeta, J. Stochel, Spectral order for unbounded operators. J. Math. Anal. Appl. **389**(2), 1029–1045 (2012)
70. C.R. Putnam, On normal operators in Hilbert space. Am. J. Math. **73**, 357–362 (1951)
71. C.R. Putnam, On square roots of normal operators. Proc. Am. Math. Soc. **8**, 768–769 (1957)
72. C.R. Putnam, in *Commutation Properties of Hilbert Space Operators and Related Topics* (Springer-Verlag, New York, 1967)
73. H. Radjavi, P. Rosenthal, On invariant subspaces and reflexive algebras. Am. J. Math. **91**, 683–692 (1969)
74. H. Radjavi, P. Rosenthal, On roots of normal operators. J. Math. Anal. Appl. **34**, 653–664 (1971)
75. M. Reed, B. Simon, Methods of modern mathematical physics, in *Functional Analysis*, vol. 1 (Academic Press, San Diego, 1980)
76. M. Reed, B. Simon, Methods of modern mathematical physics, in *Fourier Analysis, Self-Adjointness*, vol. 2 (Academic Press, New York, 1975)
77. W. Rehder, On the product of self-adjoint operators. Internat. J. Math. Math. Sci. **5**(4), 813–816 (1982)
78. W. Rudin, in *Functional Analysis*, 2nd edn. (McGraw-Hill, New York, 1991)
79. K. Schmüdgen, On commuting unbounded selfadjoint operators I. Acta Sci. Math. (Szeged) **47**(1–2), 131–146 (1984)
80. K. Schmüdgen, Unbounded operator algebras and representation theory, in *Operator Theory: Advances and Applications*, vol. 37 (Birkhäuser Verlag, Basel, 1990)
81. K. Schmüdgen, Integrable operator representations of $\mathbb{R}_q^2$, $X_{q,\gamma}$ and $SL_q(2,\mathbb{R})$. Commun. Math. Phys. **159**(2), 217–237 (1994)
82. K. Schmüdgen, Operator representations of $\mathcal{U}_q(\mathrm{sl}_2(\mathbb{R}))$. Lett. Math. Phys. **37**(2), 211–222 (1996)
83. K. Schmüdgen, in *Unbounded Self-Adjoint Operators on Hilbert Space*. Graduate Texts in Mathematics, vol. 265 (Springer, Berlin, 2012)
84. J.G. Stampfli, Roots of scalar operators. Proc. Am. Math. Soc. **13**, 796–798 (1962)
85. J.G. Stampfli, B.L. Wadhwa, An asymmetric Putnam-Fuglede theorem for dominant operators. Indiana Univ. Math. J. **25**(4), 359–365 (1976)
86. J.G. Stampfli, B.L. Wadhwa, On dominant operators. Monatsh. Math. **84**(2), 143–153 (1977)

87. J. Stochel, F.H. Szafraniec, Domination of unbounded operators and commutativity. J. Math. Soc. Jpn. **55**(2), 405–437 (2003)
88. F.-H. Vasilescu, Anticommuting self-adjoint operators. Rev. Roum. Math. Pures Appl. **28**(1), 76–91 (1983)
89. J. Weidmann, in *Linear Operators in Hilbert Spaces* (translated from the German by J. Szücs). Graduate Texts in Mathematics, vol. 68 (Springer-Verlag, 1980)
90. P.Y. Wu, All (?) about quasinormal operators, in *Operator Theory and Complex Analysis*, vol. 59, pp. 372–398. Operator Theory: Advances and Applications (Birkhäuser, Basel, 1992)
91. T. Yamazaki, M. Yanagida, Relations between two operator inequalities and their applications to paranormal operators. Acta Sci. Math. (Szeged) **69**(1–2), 377–389 (2003)
92. J. Yang, H.-K. Du, A note on commutativity up to a factor of bounded operators. Proc. Am. Math. Soc. **132**(6), 1713–1720 (2004)
93. R. Zeng, Young's inequality in compact operators-the case of equality. J. Inequal. Pure Appl. Math. **6**(4), Article 110, 10 pp. (2005)

Chapter 7
Some Other Intertwining Relations

7.1 Beck-Putnam's Theorem and Its Generalizations

W.A. Beck and C.R. Putnam proved in [2] the following result:

Theorem 7.1.1 *Let A be a bounded operator on a Hilbert space. Let N be a normal operator such that $AN = N^*A$. If, whenever z is not real, either z or its conjugate $\bar{z}$ does not lie in the spectrum of N, then $AN = NA$.*

W. Rehder [28] gave a generalization of this theorem and proved:

Theorem 7.1.2 *Assume that A is a bounded operator on a Hilbert space. Let N and M be two normal operators such that $AN = MA$. Assume further that*

1. *$\sigma(M^*) \subset \sigma(N)$, and*
2. *whenever z is not real, either z or its conjugate $\bar{z}$ does not lie in the spectrum of N.*

Then $AN = M^*A$.

W. Rehder's proof was interesting, however, not a straightforward one. In [17], Theorem 7.1.2 was reduced to Theorem 7.1.1 utilizing Berberian's trick. Incidentally, this has led to a slight weakening of the assumptions of Theorem 7.1.2. The improved version reads:

Theorem 7.1.3 ([17]) *Assume that $A, N, M \in B(H)$. Let N and M be both normal and such that $AN = MA$. Assume further that, whenever z is not real, either z or its conjugate $\bar{z}$ does not lie in $\sigma(M^*) \cup \sigma(N)$. Then $AN = M^*A$.*

Proof *([17])* Define on $H \oplus H$ the following operators

$$\tilde{N} = \begin{pmatrix} M^* & 0 \\ 0 & N \end{pmatrix} \text{ and } \tilde{A} = \begin{pmatrix} 0 & A \\ 0 & 0 \end{pmatrix}.$$

M. H. Mortad, *The Fuglede-Putnam Theory*, Lecture Notes in Mathematics 2322,
https://doi.org/10.1007/978-3-031-17782-8_7

Since M and N are normal, so is $\widetilde{N}$. In addition, one has

$$\widetilde{A}\widetilde{N} = \begin{pmatrix} 0 & AN \\ 0 & 0 \end{pmatrix} \text{ and } \widetilde{N}^*\widetilde{A} = \begin{pmatrix} 0 & MA \\ 0 & 0 \end{pmatrix}.$$

Since $AN = MA$, it is seen that $\widetilde{A}\widetilde{N} = \widetilde{N}^*\widetilde{A}$. We also know that $\sigma(\widetilde{N}) = \sigma(M^*) \cup \sigma(N)$. However, and by hypothesis, whenever z is not real either $z \notin \sigma(M^*) \cup \sigma(N)$ or $\bar{z} \notin \sigma(M^*) \cup \sigma(N)$ and hence whenever z is not real either $z \notin \sigma(\widetilde{N})$ or $\bar{z} \notin \sigma(\widetilde{N})$. Thus Theorem 7.1.1 applies and gives

$$\widetilde{A}\widetilde{N} = \widetilde{N}\widetilde{A} \text{ or } AN = M^*A$$

as wished.   □

Remark Theorem 7.1.2 has now become a corollary to the previous theorem.

Theorem 7.1.3 has an analog for unbounded operators. Its proof uses a Rosenblum-like argument:

Theorem 7.1.4 ([17]) *Assume that A is a bounded operator on a Hilbert space. Let N and M be two unbounded normal operators (with domains $D(N)$ and $D(M)$ respectively) such that $AN \subset MA$. Assume further that, whenever z is not real, either z or its conjugate $\bar{z}$ does not lie in $\sigma(M^*) \cup \sigma(N)$. Then $AN \subset M^*A$.*

Proof *([17])* Let m and n be both in $\mathbb{N}$. Consider the two closed balls

$$B_n = \{z \in \mathbb{C} : |z| \le n\} \text{ and } B_m = \{z \in \mathbb{C} : |z| \le m\}.$$

Let $P_{B_n}(N)$ and $P_{B_m}(M)$ be the spectral projections associated with N and M respectively. Then $N_n := NP_{B_n}(N)$ and $M_m := MP_{B_m}(M)$ are bounded normal operators.

As $AN \subset MA$, AN and MA coincide on $D(AN) = D(N)$. Since ran $P_{B_n}(N) \subset D(N)$, one can say that $ANP_{B_n}(N) = MAP_{B_n}(N)$.

Known properties about the spectral measures and theorem then yield:

$$[P_{B_m}(M)AP_{B_n}(N)]N_n = M_m[P_{B_m}(M)AP_{B_n}(N)].$$

Let z be a non-real number. Since $z \notin \sigma(M^*) \cup \sigma(N)$ or $\bar{z} \notin \sigma(M^*) \cup \sigma(N)$, $z \notin \sigma(M_m^*) \cup \sigma(N_n)$ or $\bar{z} \notin \sigma(M_m^*) \cup \sigma(N_n)$ for all n and all m. Theorem 7.1.3 then applies and yields

$$[P_{B_m}(M)AP_{B_n}(N)]NP_{B_n}(N) = P_{B_m}(M)M^*[P_{B_m}(M)AP_{B_n}(N)].$$

Thus

$$P_{B_m}(M)ANP_{B_n}(N)f = P_{B_m}(M)M^*AP_{B_n}(N)f, \ \forall f \in D(N) = D(N^*)$$

(and hence $Af \in D(M) = D(M^*)$). Sending both n and m to ∞ we get $P_{B_n}(N) \to I$ and $P_{B_m}(M) \to I$ respectively (both w.r.t. the strong operator topology).
    Therefore

$$ANf = M^*Af, \ \forall f \in D(N) \ (\subset D(M^*A) = D(MA)).$$

Thus $AN \subset M^*A$, as required.   □

7.2 Berberian's Theorem: Generalizations and Applications

We start with a particular definition:

Definition 7.2.1 A unitary operator U is said to be cramped if its spectrum is completely contained on some open semi-circle (of the unit circle), that is

$$\sigma(U) \subseteq \{e^{it} : \alpha < t < \alpha + \pi\}.$$

S.K. Berberian established the following result:

Theorem 7.2.1 ([4]) *If U is a cramped unitary element of $\mathcal{A}$ (where $\mathcal{A}$ is a C^*-algebra), and T is an element of $\mathcal{A}$ such that $UTU^* = T^*$, then T is self-adjoint.*

Its proof is based upon the so-called Cayley transform. S. Dehimi and M. H. Mortad then obtained the following generalization of Berberian's theorem, whose proof bypasses the Cayley transform. This obviously constitutes another proof of the bounded case.

Theorem 7.2.2 ([8]) *Let U be a cramped unitary operator. Let T be a closed and densely defined operator such that $UT = T^*U$. Then T is self-adjoint.*

Proof *([8])* First we prove that $U^2T = TU^2$. Since U is bounded and invertible, we have

$$(UT)^* = T^*U^* \text{ and } (T^*U)^* = U^*T^{**} = U^*\overline{T} = U^*T.$$

Hence $T^*U^* = U^*T$. We may then write

$$U^2TU^{*2} = U(UTU^*)U^* = UT^*U^* = UU^*T = T,$$

giving $U^2T = TU^2$ or $TU^{*2} = U^{*2}T$ or $U^2T^* = T^*U^2$.

Next, we prove that $TU = UT^*$. We have

$$TU = U^*T^*UU = U^*T^*U^2 = U^*UUT^* = UT^*.$$

Hence also $U^*T^* = TU^*$.

The penultimate step in the proof is to prove that T is normal. To this end, set $S = \frac{1}{2}(U + U^*)$. Following [36], $S > 0$.

Then we show that $STT^* \subset T^*TS$. We have

$$UTT^* = T^*UT^* = T^*TU \text{ and } U^*TT^* = T^*U^*T^* = T^*TU^*.$$

Hence

$$STT^* = \frac{1}{2}(U + U^*)TT^*$$

$$= \frac{1}{2}UTT^* + \frac{1}{2}U^*TT^* \text{ (as } U \text{ is bounded)}$$

$$= \frac{1}{2}T^*TU + \frac{1}{2}T^*TU^*$$

$$\subset T^*TS.$$

So, according to Corollary 5.1 in [34], $TT^* = T^*T$. Since T is closed, we immediately conclude that T is normal. Accordingly, and by Corollary 3 in [18],

$$UT = T^*U \Longrightarrow T = T^*$$

as $0 \notin \overline{W(U)}$, establishing the result.   □

Remark A hypothesis like $UT \subset T^*U$ would not yield the desired result. For example, take T to be any symmetric and closed unbounded operator T that is not self-adjoint. Let $U = I$ be the identity operator on the given Hilbert space. Then clearly $UT \subset T^*U$ while $T \neq T^*$.

The assumption that U being cramped is indispensable even in the bounded case. This was already observed by W. A. Beck and C. R. Putnam [2], and C. A. Mc-Carthy [15].

We want to close this section with applications of Berberian's theorem to exponentials of normal operators. The following lemmata, whose proofs are based upon the functional calculus, will come in handy for the proofs of the results below.

Lemma 7.2.3 *Let A and B be two commuting normal operators, on a Hilbert space, having spectra contained in simply connected regions not containing 0. Then*

$$A^i B^i = B^i A^i$$

where $i = \sqrt{-1}$ is the usual complex number.

Lemma 7.2.4 *Let A be a self-adjoint operator such that $\sigma(A) \subset (0, \pi)$. Then*

$$(e^{iA})^i = e^{-A}.$$

The following result is well-known.

Theorem 7.2.5 *Let $A, B \in B(H)$ be self-adjoint. Then*

$$e^A e^B = e^B e^A \Longleftrightarrow AB = BA.$$

We include a simple proof for ease of reference.

Proof (More details may be found in [20].) We only show the implication "$\Rightarrow$". Suppose $e^A e^B = e^{A+B}$. Since $A + B$ is self-adjoint, e^{A+B} too is self-adjoint. Hence so is $e^A e^B$, i.e.,

$$e^A e^B = (e^A e^B)^* = (e^B)^*(e^A)^* = e^B e^A.$$

Since e^A is positive, we may write

$$f(e^A)e^B = e^B f(e^A) \text{ or merely } \ln(e^A)e^B = e^B \ln(e^A)$$

and so

$$Ae^B = e^B A.$$

Since B is self-adjoint too, the same reasoning mutatis mutandis gives $AB = BA$, as needed. $\square$

One may ask: To what extent does the previous result hold for normal operators with a bare minimum of assumptions? Several mathematicians have generalized Theorem 7.2.5 using the so-called $2\pi i$-congruence-free hypothesis on spectra (see, e.g., [26, 29, 35]). Here we present a different and original way of tackling such problems. The main idea is to try to get back to Theorem 7.2.5, using the Cartesian decomposition.

We begin with the case where one operator is normal while the other one remains self-adjoint:

Proposition 7.2.6 ([19]) *Let N be a normal operator with Cartesian decomposition $A + iB$. Let S be a self-adjoint operator. If $\sigma(B) \subset (0, \pi)$, then*

$$e^S e^N = e^N e^S \Longrightarrow SN = NS.$$

Proof *([19])* Let $N = A + iB$ where A and B are two commuting self-adjoint operators. Hence $e^A e^{iB} = e^{iB} e^A$. Consequently,

$$e^S e^N = e^N e^S \iff e^S e^A e^{iB} = e^A e^{iB} e^S$$

$$\iff e^S e^A e^{iB} = e^{iB} e^A e^S$$

$$\iff e^S e^A e^{iB} = e^{iB} (e^S e^A)^*$$

Since B is self-adjoint, e^{iB} is unitary. It is in fact cramped thanks to the spectral hypothesis on B. Now, Theorem 7.2.1 implies that $e^S e^A$ is self-adjoint, i.e., $e^S e^A = e^A e^S$, so that Theorem 7.2.5 gives $AS = SA$.

It only remains to show that $BS = SB$. Since $e^S e^A = e^A e^S$, we obtain

$$e^S e^N = e^N e^S \implies e^S e^A e^{iB} = e^A e^{iB} e^S \text{ or } e^A e^S e^{iB} = e^A e^{iB} e^S,$$

and so $e^S e^{iB} = e^{iB} e^S$ by the invertibility of e^A.

Using Lemmas 7.2.5 and 7.2.4, it is immediately seen that

$$e^S e^{-B} = e^{-B} e^S.$$

Consequently, Theorem 7.2.5 yields $BS = SB$, whereby

$$SN = S(A + iB) = (A + iB)S = NS,$$

and this marks the end of the proof.   □

Remark By looking closely at the proof of Proposition 7.2.6, we see that we may well take $(-\frac{\pi}{2}, \frac{\pi}{2})$ instead of $(0, \pi)$ without altering the conclusion of the proposition. Hence the same results hold with this new interval. Thus any self-adjoint operator obeys the given condition on the spectrum.

Next, we state and prove the result involving two bounded normal operators.

Theorem 7.2.7 ([19]) *Let N and M be two normal operators with Cartesian decompositions $A + iB$ and $C + iD$ respectively. If $\sigma(B), \sigma(D) \subset (0, \pi)$, then*

$$e^M e^N = e^N e^M \implies MN = NM.$$

Proof *([19])* Since M is normal, e^M is normal too. So

$$e^M e^N = e^N e^M \implies e^{M^*} e^N = e^N e^{M^*}$$

by the Fuglede theorem. Using again the normality of M, we have

$$e^{M^*} e^M e^N = e^{M^*} e^N e^M \implies e^{M^*} e^M e^N = e^N e^{M^*} e^M$$

or merely

$$e^{M^*+M}e^N = e^N e^{M^*+M}$$

Since $M^* + M$ is self-adjoint, Proposition 7.2.6 applies and gives

$$(M^* + M)N = N(M^* + M) \text{ or just } CN = NC.$$

Therefore, $N^*C = CN^*$ and so $(N + N^*)C = C(N + N^*)$. Thus

$$AC = CA \text{ and hence } BC = CB.$$

Similar arguments and a glance at Proposition 7.2.6 all yield

$$AM = MA \text{ and hence } AD = DA.$$

To prove the remaining bit, we go back to the equation $e^N e^M = e^M e^N$. Since $BC = CB$ and $AD = DA$, we obtain

$$e^A e^{iB} e^C e^{iD} = e^C e^{iD} e^A e^{iB} \iff e^A e^C e^{iB} e^{iD} = e^C e^A e^{iD} e^{iB}.$$

Since A and C commute and since $e^A e^C$ is invertible, we are left with

$$e^{iB} e^{iD} = e^{iD} e^{iB}.$$

Lemmas 7.2.3 and 7.2.4 then yield

$$e^{-B} e^{-D} = e^{-D} e^{-B}$$

which, in turn, leads to $BD = DB$. Hence $BM = MB$.
Finally, we have

$$NM = (A + iB)M = AM + iBM = MA + iMB = M(A + iB) = MN,$$

and the proof is complete.   □

From the above results, we may derive quite a few simple but interested consequences. First, we need the coming result:

Theorem 7.2.8 ([6]) *Let A be in $B(H)$. Let $N \in B(H)$ be normal and such that $\sigma(\operatorname{Im} N) \subset (0, \pi)$. Then*

$$Ae^N = e^N A \iff AN = NA.$$

Proof *([6])* Obviously, we are only concerned with proving the implication "⇒". The normality of N implies that of e^N, and so by the Fuglede theorem

$$Ae^N = e^N A \Longleftrightarrow Ae^{N^*} = e^{N^*} A.$$

Hence

$$A^* e^N = e^N A^*,$$

and so

$$(A + A^*)e^N = e^N(A + A^*) \text{ or } (\mathrm{Re}A)e^N = e^N(\mathrm{Re}A)$$

so that

$$e^{\mathrm{Re}\,A} e^N = e^N e^{\mathrm{Re}A}.$$

Since $\mathrm{Re}\,A$ is self-adjoint, Proposition 7.2.6 applies and gives

$$(\mathrm{Re}\,A)N = N(\mathrm{Re}\,A).$$

Similarly, we find that

$$(\mathrm{Im}\,A)e^N = e^N(\mathrm{Im}\,A).$$

Taking into account the self-adjointness of $\mathrm{Im}\,A$, similar arguments to those applied before then yield

$$(\mathrm{Im}\,A)N = N(\mathrm{Im}\,A).$$

Therefore, $AN = NA$, as required.   □

Remark See [6] for a counterexample that shows the importance of the hypothesis $\sigma(\mathrm{Im}\,N) \subset (0, \pi)$ for the result to hold.

The first consequence of the previous result reads:

Theorem 7.2.9 ([6]) *Let A and B be both in $B(H)$. Assume that $A + B$ is normal and such that $\sigma(\mathrm{Im}(A + B)) \subset (0, \pi)$. If*

$$e^A e^B = e^B e^A = e^{A+B},$$

then $AB = BA$.

Proof *([6])* Write

$$e^{A+B}e^A = e^Be^Ae^A = e^Ae^Be^A = e^Ae^{A+B}.$$

Since $A + B$ is normal and $\sigma(\text{Im}(A + B)) \subset (0, \pi)$, Theorem 7.2.8 gives

$$(A + B)e^A = e^A(A + B) \text{ or just } Be^A = e^AB$$

for A patently commutes with e^A. Now, right multiplying both sides of the previous equation by e^B leads to

$$Be^Ae^B = e^ABe^B = e^Ae^BB$$

or

$$Be^{A+B} = e^{A+B}B.$$

Applying again Theorem 7.2.8, we see that

$$B(A + B) = (A + B)B \text{ or } AB = BA.$$

The proof is thus complete. $\square$

Corollary 7.2.10 *Let $A \in B(H)$. Then*

$$e^Ae^{A^*} = e^{A^*}e^A = e^{A+A^*} \iff A \text{ is normal.}$$

Proof We need only prove the implication "$\Rightarrow$". It is plain that $A + A^*$ is self-adjoint. Hence the remark below the proof of Proposition 7.2.6 combined with Theorem 7.2.9 give us

$$AA^* = A^*A.$$

$\square$

We have yet another consequence of Theorem 7.2.8 (cf. [30]).

Corollary 7.2.11 ([6]) *Let A be normal such that $\sigma(\text{Im } A) \subset (0, \pi)$. Let $B \in B(H)$. Then*

$$e^A = e^B \implies A^2B = BA^2.$$

Before carrying on, recall the following useful result (a proof can be consulted in, e.g., Exercise 11.3.14 in [20]):

Lemma 7.2.12 *Let $T \in B(H)$ be self-adjoint and such that $e^T = I$, where I designates the identity operator on H. Then $T = 0$.*

The following known result is usually shown using the spectral theorem.

Corollary 7.2.13 *Let $A \in B(H)$ be normal. Then*

$$A \text{ is self-adjoint} \iff e^{iA} \text{ is unitary.}$$

Proof The implication "$\Rightarrow$" is well-known even without using the spectral theorem (see, e.g., Exercise 4.3.33 in [20]). Let us prove the reverse implication. By the normality of A, we have

$$e^{iA - iA^*} = e^{iA} e^{-iA^*} = e^{iA} (e^{iA})^* = I.$$

Since $iA - iA^*$ is self-adjoint, the preceding equation and Lemma 7.2.12 give $A = A^*$, and this completes the proof. $\square$

We now come to a result that appeared in [26] and [29]. It reads: If A is self-adjoint, $\sigma(A) \subseteq [-\pi, \pi]$ and $e^{iA} = e^B$, then $B^* = -B$ if B is normal. Here is an improvement of this result.

Proposition 7.2.14 ([6]) *If A is self-adjoint and $e^{iA} = e^B$, then $B^* = -B$ whenever B is normal.*

Proof ([6]) It is clear that e^{iA} is unitary. We also have

$$e^{B^*} = e^{-iA} \text{ and } e^{-B} = e^{-iA}.$$

Thus

$$e^{-B} = e^{B^*} \text{ so that } e^{B + B^*} = I$$

because B is normal. However, $B + B^*$ is always self-adjoint, whence $B^* = -B$ by Lemma 7.2.12. $\square$

Remark Readers interested in this topic may consult, e.g., [6, 19, 20, 22, 27, 29], and [30] for further reading.

7.3 Sheth-Williams' Theorem and Its Generalization

I.H. Sheth proved the following result:

Theorem 7.3.1 ([33]) *Let T be a bounded hyponormal operator. If S is any bounded operator for which $0 \notin \overline{W(S)}$, then*

$$ST = T^*S \Longrightarrow T = T^*.$$

Then, J.P. Williams [36] obtained a much better conclusion in the previous theorem using the technique of [33]. More precisely, he obtained the realness of the spectrum of T.

Now, we extend Theorem 7.3.1 to unbounded operators. The proof requires the following two lemmata:

Proposition 7.3.2 ([13]) *Let T be an unbounded, closed and hyponormal operator in some Hilbert space H. Then $W(T) \subset \operatorname{conv} \sigma(T)$, where $\operatorname{conv} \sigma(T)$ denotes the the convex hull of $\sigma(T)$.*

Proposition 7.3.3 ([25]) *Let T be a densely defined, closed and symmetric operator in a Hilbert space. If T is quasi-similar to its adjoint T^*, then T is self-adjoint (for the definition of quasi-similarity, the reader may look at [25]).*

Here is the promised generalization:

Theorem 7.3.4 ([8]) *Let S be a bounded operator on a $\mathbb{C}$-Hilbert space H such that $0 \notin \overline{W(S)}$. Let T be an unbounded and closed hyponormal operator with a dense domain $D(T) \subset H$. If $ST^* \subset TS$, then T is self-adjoint.*

Proof *([8])* The proof is divided into three claims:

1. **Claim 1:** $\sigma_a(T^*) = \sigma(T^*)$ (where σ_a designates the approximate point spectrum). By definition, $\sigma_a(T^*) \subset \sigma(T^*)$. To show the reverse inclusion, let $\lambda \notin \sigma_a(T^*)$. Then there exists a positive number k such that

$$\|T^*x - \lambda x\| \geq k\|x\| \text{ for all } x \in D(T^*).$$

Hence $T^* - \lambda I$ is clearly injective. Besides, $\operatorname{ran}(T^* - \lambda I)$ is closed as $T^* - \lambda I$ is closed for T^* is so. Now, since T is hyponormal, so is $T - \bar\lambda I$. This signifies that

$$\|Tx - \bar\lambda x\| \geq \|T^*x - \lambda x\| \geq k\|x\| \text{ for all } x \in D(T) \subset D(T^*).$$

Whence $T - \bar\lambda I$ is also injective so that

$$\operatorname{ran}(T^* - \lambda I)^\perp = \ker(T - \bar\lambda I) = \{0\} \text{ or } \overline{\operatorname{ran}(T^* - \lambda I)} = H.$$

Thus $T^* - \lambda I$ is onto since we already observed that its range was closed. Therefore, $\lambda \notin \sigma(T^*)$.

2. **Claim 2:** $\sigma(T) \subset \mathbb{R}$. Let $\lambda \in \sigma(T^*) = \sigma_a(T^*)$. Then for some $x_n \in D(T^*)$ such that $\|x_n\| = 1$ we have $\|T^*x_n - \lambda x_n\| \to 0$ as $n \to \infty$. Since $ST^* \subset TS$ and $x_n \in D(T^*)$, we have $ST^*x_n = TSx_n$ so that we may write the following

$$
\begin{aligned}
0 \le |(\bar{\lambda} - \lambda)\langle Sx_n, x_n \rangle| &= |\langle (ST^*S^{-1} - \lambda + \bar{\lambda} - T)Sx_n, x_n \rangle| \\
&\le |\langle (S(T^* - \lambda I)x_n, x_n \rangle| + |\langle (\bar{\lambda} - T)Sx_n, x_n \rangle| \\
&\le \|S\| \, \|T^*x_n - \lambda x_n\| + |\langle Sx_n, (\lambda - T^*)x_n \rangle| \\
&\le \|S\| \, \|T^*x_n - \lambda x_n\| + \|S\| \, \|T^*x_n - \lambda x_n\| \\
&= 2\|S\| \, \|T^*x_n - \lambda x_n\| \to 0.
\end{aligned}
$$

(where in the second inequality we used the fact that $x_n \in D(T^*)$ and $Sx_n \in D(T)$ both coming from $ST^* \subset TS$). However, the condition $0 \notin W(S)$ forces us to have $\lambda = \bar{\lambda}$. Accordingly, $\sigma(T^*) \subset \mathbb{R}$ or just $\sigma(T) \subset \mathbb{R}$ (remember that $\sigma(T^*) = \{\bar{\lambda} : \lambda \in \sigma(T)\}$).

3. **Claim 3:** $T = T^*$. Since $\sigma(T) \subset \mathbb{R}$, Proposition 7.3.2 implies that $W(T) \subset \mathbb{R}$, which clearly implies that $\langle Tx, x \rangle \in \mathbb{R}$ for all $x \in D(T)$ which, in its turn, means that T is symmetric. Hence T is quasi-similar to T^* via S and I, so that Proposition 7.3.3 applies and gives the self-adjointness of T. This completes the proof.

□

Remark ([8]) The condition $ST^* \subset TS$ in the previous result is not purely conventional. In other words, we cannot obtain the desired result by simply assuming that $ST \subset T^*S$ instead, even with a slightly stronger condition (namely symmetricity in lieu of hyponormality).

For example, consider an unbounded, densely defined, closed, and symmetric operator T that is *not self-adjoint*, i.e., $T \ne T^*$. Let $S = I$ be the identity operator on H. Then S is bounded and $0 \notin \overline{W(S)}$. Also, T is closed and hyponormal. Finally, it is plain that

$$T = ST \subset T^* = T^*S.$$

Nonetheless, we can still obtain the self-adjointness of T out of $ST \subset T^*S$ by imposing an extra condition on T.

Theorem 7.3.5 ([8]) *Let $S \in B(H)$ be such that $0 \notin \overline{W(S)}$. Let T be an unbounded hyponormal operator that is also boundedly invertible. If $ST \subset T^*S$, then T is self-adjoint.*

Proof *([8])* Since T is invertible with an everywhere defined bounded inverse, we have

$$ST \subset T^*S \Longrightarrow ST^{-1} = (T^{-1})^*S.$$

Since T is hyponormal, the bounded T^{-1} too is hyponormal (see [12]). Hence by [33], T^{-1} is self-adjoint. Hence

$$T^{-1} = (T^{-1})^* \Longrightarrow T(T^{-1})^* = I \Longrightarrow T^* \subset T.$$

Finally, since T is hyponormal, $D(T) \subset D(T^*)$ and so $T = T^*$, as suggested. $\square$

7.4 Operator Equations Involving the Shift Operator

In this section, S will denote the usual unilateral shift operator, i.e., $Se_n = e_{n+1}$ for $n \geq 0$, where (e_n) is some orthonormal basis in H.

Theorem 7.4.1 ([14]) *If A is hyponormal and V is an isometry, and if X is an operator such that $AX = XV^*$, then $\|Xf\| = \|XV^n f\| = \|XV^{*n}\|$ for all $f \in H$ and $n \geq 0$.*

Corollary 7.4.2 ([14]) *Let $A \in B(H)$ be hyponormal. If $X \in B(H)$ is such that $AX = XS^*$, then $X = 0$.*

Proof As is known, $S^{*n} \to 0$ strongly, as $n \to \infty$. By Theorem 7.4.1, we infer that $X = 0$. $\square$

Corollary 7.4.3 ([14]) *Let $A \in B(H)$ be such that $\ker A \subset \ker A^*$. If $X \in B(H)$ is such that $AX = XS^*$, then $X = 0$.*

Proof Let $n \in \mathbb{Z}^+$. Then, $AXe_{n+1} = Xe_n$ and $AXe_0 = 0$. We claim that $A^2Xe_{n+1} = 0$ for $n \geq 0$. The statement is true for $n = 0$ for $A^2Xe_1 = AXe_0 = 0$. Now, suppose that $A^2Xe_n = 0$. By the assumption $\ker A \subset \ker A^*$, it ensues that $A^*AXe_n = 0$. Hence

$$\|AXe_n\|^2 = \langle A^*AXe_n, Xe_n \rangle = 0,$$

i.e., $AXe_n = 0$. The equation $AXe_{n+1} = Xe_n$ then yields $A^2Xe_{n+1} = 0$. Hence $A^*AXe_{n+1} = 0$. Therefore,

$$\|Xe_n\|^2 = \langle AXe_{n+1}, AXe_{n+1} \rangle = \langle A^*AXe_{n+1}, Xe_n \rangle = 0,$$

and so $Xe_n = 0$ for all n. In other language, $X = 0$, as needed. $\square$

Remark Notice that all the above results were in fact established for A being "M-hyponormal" instead of "hyponormal". Other related results may be consulted in [14].

The ensuing result appeared in [16]:

Corollary 7.4.4 *Assume X is such that $0 \notin W(X)$ (or that $\sigma(X) \cap \sigma(-X) = \emptyset$). Let S be the unilateral shift. Then the operator equation*

$$XS = S^*X$$

has no non-zero normal solution X on ℓ^2.

Remark In the end, readers may find other intertwining relations in, e.g., [11] and [32], and further bibliography cited therein.

7.5   Counterexamples

We start with some counterexamples on finite-dimensional spaces.

Suppose $B \in B(H)$ is normal and $A \in B(H)$ is self-adjoint. Do we have

$$BA = AB \implies B^*A = AB? \text{ Or } BA = AB^*?$$

The answer is negative, i.e., both implications are false.

Example 7.5.1 The same counterexample works for either of them. Consider a non-self-adjoint normal operator $B \in B(H)$ and take $A = I$. Then plainly $BA = AB$, but neither $B^*A = AB$ nor $BA = AB^*$ needs to hold.

The next example is related to Theorem 7.1.1.

Example 7.5.2 ([2]) Let

$$A = \begin{pmatrix} 1 & 0 \\ 0 & -1 \end{pmatrix} \text{ and } U = \begin{pmatrix} 0 & -1 \\ 1 & 0 \end{pmatrix}.$$

Then A is self-adjoint and U is unitary. Hence,

$$UA = \begin{pmatrix} 0 & 1 \\ 1 & 0 \end{pmatrix}, \; AU^* = \begin{pmatrix} 0 & 1 \\ 1 & 0 \end{pmatrix} \text{ and } AU = \begin{pmatrix} 0 & -1 \\ -1 & 0 \end{pmatrix}$$

and so

$$AU^* = UA \text{ and } AU \neq UA,$$

as needed.

The following example shows that there is a closed operator T in some Hilbert space, which does not commute with any bounded operator $S \in B(H)$, except for scalar operators (i.e., those of the type αI for some scalar α).

Example 7.5.3 ([10], cf. [24]) Let A be the multiplication operator by x and let $B = -id/dx$, both defined on their maximal domains $D(A)$ and $D(B)$ in $L^2(\mathbb{R})$ respectively (recall that A and B are self-adjoint). Define now the operator $T :=$ $A + iB$ by

$$Tf(x) = xf(x) + f'(x)$$

for every $f \in D := D(T) = \{f \in L^2(\mathbb{R}) : xf, f' \in L^2(\mathbb{R})\}$. It is known that T is closed (see, e.g., [10]).

If $\lambda \in \mathbb{C}$, then the equation $Tf = \lambda f$ may be expressed as

$$f'(x) + (x - \lambda)f(x) = 0.$$

It is known that the only solutions of the previous equation are given by $f(x) = Kf_\lambda(x)$ where K is a constant and

$$f_\lambda(x) = e^{-\frac{(x-\lambda)^2}{2}}$$

(observe that $f_\lambda \in D$ for any λ). Hence λ is a simple eigenvalue for T with cf_λ being the associated eigenvector (where $c \in \mathbb{C}$).

Assume now for the sake of contradiction that T commutes with an everywhere defined bounded operator S, that is, $ST \subset TS$. This means that $STf = TSf$ for all $f \in D(T) \subset D(TS)$. In particular,

$$TSf_\lambda = \lambda Sf_\lambda, \; \forall \lambda \in \mathbb{C}.$$

This actually signifies that λ is an eigenvalue for T with Sf_λ being an eigenvector (in case $Sf_\lambda \neq 0$). Therefore, we may write $Sf_\lambda = c_\lambda f_\lambda$ where c_λ is some complex number depending on λ only.

Let's show that c_λ is in fact a differentiable function of λ. Since c_λ is bounded (as S is), we deduce by Liouville's theorem that c_λ is a constant function, that is, c_λ becomes just a constant c. First, it may be shown (in H) that

$$\frac{f_\lambda - f_\mu}{\lambda - \mu} \longrightarrow g_\mu, \quad \text{when } \lambda \to \mu.$$

As a consequence, if h is any fixed element of $L^2(\mathbb{R})$, then the complex function $\lambda \mapsto \langle f_\lambda, h \rangle$ is differentiable at any λ. The objective now is to show that c_λ is differentiable at *any* given point μ. Since $Sf_\lambda = c_\lambda f_\lambda$, it follows that

$$c_\lambda \langle f_\lambda, f_\mu \rangle = \langle c_\lambda f_\lambda, f_\mu \rangle = \langle Sf_\lambda, f_\mu \rangle = \langle f_\lambda, S^* f_\mu \rangle.$$

Here both $\langle f_\lambda, f_\mu \rangle$ and $\langle f_\lambda, S^* f_\mu \rangle$ are differentiable. Therefore,

$$c_\lambda = \frac{\langle f_\lambda, S^* f_\mu \rangle}{\langle f_\lambda, f_\mu \rangle}$$

is differentiable at any point λ provided that $\langle f_\lambda, f_\mu \rangle \neq 0$. This is particularly the case at $\lambda = \mu$. Finally, as $Sf_\lambda = cf_\lambda$ for all λ, we infer that

$$Sf = cf, \ \forall f \in L^2(\mathbb{R})$$

given that S is bounded and that the set of all finite linear combinations of the elements f_λ is everywhere dense in $L^2(\mathbb{R})$.

Next, we give examples in the same spirit. First, we say a few words about a striking example by P. R. Chernoff, who found a simpler example in [7] than the pioneering one due to M. Naimark in [23]. P.R. Chernoff found an explicit example of a closed, unbounded, densely defined, symmetric and positive operator A (defined in the Hardy space $H^2(S)$, where S is the unit circle) such that $D(A^2) = \{0\}$. It is worth noticing that K. Schmüdgen showed in [31] (simultaneously as P. R. Chernoff) that every unbounded self-adjoint T has two closed and symmetric restrictions A and B such that

$$D(A) \cap D(B) = \{0\} \text{ and } D(A^2) = D(B^2) = \{0\}.$$

The latter result was then generalized later by J.F. Brasche and H. Neidhardt in [5] (see also [1]). Recently, S. Dehimi and M.H. Mortad obtained in [9] a densely defined, one-to-one and closed operator B such that

$$D(B^2) = D(B^{*2}) = \{0\}.$$

See [21] for similar results for higher powers.

We now show the existence of a self-adjoint operator A and a densely defined closed operator B that are not intertwined by any (everywhere defined) bounded operator except the zero operator. Also, the same pair A and B in the opposite order cannot be intertwined by any bounded operator except the zero operator.

Example 7.5.4 ([22]) Let $H = L^2(\mathbb{R}) \oplus L^2(\mathbb{R})$ and let A be any unbounded self-adjoint operator with domain $D(A) \subset H$ and let B be a closed operator such that $D(B^2) = \{0\} = D(B^{*2})$. Let $T \in B(H)$. Then,

$$TA \subset BT \implies TA^2 \subset BTA \subset B^2T.$$

Hence

$$D(TA^2) = D(A^2) \subset D(B^2T) = \{x \in H : Tx \in D(B^2) = \{0\}\} = \ker T.$$

Since A^2 is densely defined, it follows that

$$H = \overline{D(A^2)} \subset \overline{\ker T} = \ker T \subset H.$$

Hence $\ker T = H$, that is, $T = 0$, as required.

Now, we pass to the second part of the question. Plainly,

$$SB \subset AS \Longrightarrow S^*A \subset B^*S^*.$$

As before, we obtain

$$S^*A^2 \subset B^{*2}S^*.$$

Similar arguments as above then yield $S^* = 0$ or merely $S = 0$, as needed.

Now, there are a self-adjoint operator A and a densely defined closed symmetric operator B (with $B \subset A$) that are not intertwined by any (everywhere defined) bounded operator except the zero operator:

Example 7.5.5 Take any unbounded self-adjoint operator A, then consider either of its two closed symmetric restrictions and denote it by B, where also $D(B^2) = \{0\}$. Finally, consider $T \in B(H)$ such that $TA \subset BT$. Then obtain $T = 0$, as done above.

Finally, there are two densely defined closed operators A and B that are not intertwined by any densely defined closed operator apart from the zero operator.

Example 7.5.6 ([3]) Let $H = L^2(\mathbb{R}) \oplus L^2(\mathbb{R})$ and let A be a densely defined closed operator with domain $D(A) \subset L^2(\mathbb{R}) \oplus L^2(\mathbb{R})$ such that $A^2 = 0$ on $D(A^2) = D(A)$ (an explicit example would be $A = \begin{pmatrix} 0 & C \\ 0 & 0 \end{pmatrix}$ defined on $H \oplus D(C)$ where C is a closed operator).

Next, let B be a closed operator satisfying $D(B^2) = \{0\}$. We have

$$TA \subset BT \Longrightarrow TA^2 \subset B^2T.$$

But

$$D(TA^2) = \{x \in D(A^2) : 0 \in D(T)\} = D(A) \text{ and } D(B^2T) = \ker T.$$

Hence

$$D(A) \subset \ker T \subset L^2(\mathbb{R}) \oplus L^2(\mathbb{R})$$

and so upon passing to the closure (w.r.t. $L^2(\mathbb{R}) \oplus L^2(\mathbb{R})$)

$$\ker T = L^2(\mathbb{R}) \oplus L^2(\mathbb{R})$$

because $\ker T$ is closed as T is closed. Therefore, $Tx = 0$ for all $x \in D(T)$, i.e., $T \subset 0$. Accordingly, as T is bounded on $D(T)$ and also closed, then $D(T)$ becomes closed and so $D(T) = H$, that is, $T = 0$ everywhere.

References

1. Y. Arlinskiĭ, V.A. Zagrebnov, Around the van Daele-Schmüdgen theorem. Integr. Equ. Oper. Theor. **81/1**, 53–95 (2015)
2. W.A. Beck, C.R. Putnam, A note on normal operators and their adjoints. J. Lond. Math. Soc. **31**, 213–216 (1956)
3. I.F.Z. Bensaid, S. Dehimi, B. Fuglede, M.H. Mortad, The Fuglede theorem and some intertwining relations. Adv. Oper. Theor. **6/1**, 8 pp. (2021). Paper No. 9
4. S.K. Berberian, A note on operators unitarily equivalent to their adjoints. J. Lond. Math. Soc. **37**, 403–404 (1962)
5. J.F. Brasche, H. Neidhardt, Has every symmetric operator a closed symmetric restriction whose square has a trivial domain? Acta Sci. Math. (Szeged) **58/1–4**, 425–430 (1993)
6. A. Chaban, M.H. Mortad, Exponentials of bounded normal operators. Colloq. Math. **133/2**, 237–244 (2013)
7. P.R. Chernoff, A semibounded closed symmetric operator whose square has trivial domain. Proc. Am. Math. Soc. **89/2**, 289–290 (1983)
8. S. Dehimi, M.H. Mortad, Bounded and unbounded operators similar to their adjoints. Bull. Korean Math. Soc. **54/1**, 215–223 (2017)
9. S. Dehimi, M.H. Mortad, Chernoff-like counterexamples related to unbounded operators. Kyushu J. Math. **74/1**, 105–108 (2020)
10. B. Fuglede, A commutativity theorem for normal operators. Proc. Natl. Acad. Sci. **36**, 35–40 (1950)
11. S. Hassi, Z. Sebestyén, H.S.V. de Snoo, On the nonnegativity of operator products. Acta Math. Hungar. **109/1–2**, 1–14 (2005)
12. J. Janas, On unbounded hyponormal operators. Ark. Mat. **27/2**, 273–281 (1989)
13. J. Janas, On unbounded hyponormal operators III. Studia Math. **112/1**, 75–82 (1994)
14. F. Kittaneh, On the operator equation $AX = XV^*$. Chin. J. Math. **16/1**, 29–40 (1988)
15. C.A. McCarthy, On a theorem of Beck and Putnam. J. Lond. Math. Soc. **39**, 288–290 (1964)
16. M.H. Mortad, Yet more versions of the Fuglede-Putnam theorem. Glasgow Math. J. **51/3**, 473–480 (2009)
17. M.H. Mortad, On a Beck-Putnam-Rehder theorem. Bull. Belg. Math. Soc. Simon Stevin **17/4**, 737–740 (2010)
18. M.H. Mortad, Similarities involving unbounded normal operators. Tsukuba J. Math. **34/1**, 129–136 (2010)
19. M.H. Mortad, Exponentials of normal operators and commutativity of operators: a new approach. Colloq. Math. **125/1**, 1–6 (2011)
20. M.H. Mortad, *An Operator Theory Problem Book* (World Scientific Publishing Co., Singapore, 2018)
21. M.H. Mortad, On the triviality of domains of powers and adjoints of closed operators. Acta Sci. Math. (Szeged) **85**, 651–658 (2019)
22. M.H. Mortad, *Counterexamples in Operator Theory* (Birkhäuser/Springer, Cham, 2022)

23. M. Naimark, On the square of a closed symmetric operator. Dokl. Akad. Nauk SSSR **26**, 866–870 (1940); ibid. **28**, 207–208 (1940)
24. E.A. Nordgren, Closed operators commuting with a weighted shift. Proc. Am. Math. Soc. **24**, 424–428 (1970)
25. S. Ôta, K. Schmüdgen, On some classes of unbounded operators. Integr. Equ. Oper. Theor. **12/2**, 211–226 (1989)
26. F.C. Paliogiannis, On commuting operator exponentials. Proc. Am. Math. Soc. **131/12**, 3777–3781 (2003)
27. F.C. Paliogiannis, On commuting operator exponentials, II. Proc. Indian Acad. Sci. Math. Sci. **123/1**, 27–31 (2013)
28. W. Rehder, On the adjoints of normal operators. Arch. Math. (Basel) **37/2**, 169–172 (1981)
29. C. Schmoeger, On normal operator exponentials. Proc. Am. Math. Soc. **130/3**, 697–702 (2001)
30. C. Schmoeger, On the operator equation $e^A = e^B$. Linear Algebra Appl. **359**, 169–179 (2003)
31. K. Schmüdgen, On domains of powers of closed symmetric operators. J. Oper. Theor. **9/1**, 53–75 (1983)
32. Z. Sebestyén, Positivity of operator products. Acta Sci. Math. (Szeged) **66/1–2**, 287–294 (2000)
33. I.H. Sheth, On hyponormal operators. Proc. Am. Math. Soc. **17**, 998–1000 (1966)
34. J. Stochel, An asymmetric Putnam-Fuglede theorem for unbounded operators. Proc. Am. Math. Soc. **129/8**, 2261–2271 (2001)
35. E.M.E. Wermuth, A remark on commuting operator exponentials. Proc. Am. Math. Soc. **125/6**, 1685–1688 (1997)
36. J.P. Williams, Operators similar to their adjoints. Proc. Am. Math. Soc. **20**, 121–123 (1969)

Chapter 8
Conjectures

In this short chapter, we list a few conjectures that could stimulate further research.

8.1 Conjectures

1. To the best of my knowledge, there isn't any unbounded version of Weiss'
   theorem. One plausible generalization is:

   Conjecture 8.1.1 Let M_1 and N_2 be two linear operators, and let $A, N_1, M_2 \in$
   $B(H)$, where M_1, M_2, N_1 and N_2 are all normal. Assume further that $N_1 N_2 \subset$
   $N_2 N_1$ and that $M_2 M_1 \subset M_1 M_2$. Then

   $$N_1 A M_1 \subset N_2 A M_2 \Longrightarrow N_1^* A M_1^* \subset N_2^* A M_2^*.$$

   So, if this conjecture is ever shown to be true, setting $N_1 = M_2 = I$ brings
   us back to the classical unbounded version of Fuglede-Putnam. Readers may try
   other generalizations.
2. It is natural to conjecture that:

   Conjecture 8.1.2 (Cf. [3, 5, 8], and [11]) Let A be a closed operator with domain
   $D(A)$. Let N be an unbounded normal operator with domain $D(N)$. If $D(N) \subset$
   $D(A)$, then

   $$AN \subset NA \Longrightarrow AN^* \subset N^* A.$$

   Notice that a "wrong proof" of the previous question appeared in the preprint
   [9]. Why was it wrong? The reason is that I thought that the assumption of the
   symmetricity of A in Theorem 5.1.4 was superfluous, while it was crucial.

© The Author(s), under exclusive license to Springer Nature Switzerland AG 2022   139
M. H. Mortad, *The Fuglede-Putnam Theory*, Lecture Notes in Mathematics 2322,
https://doi.org/10.1007/978-3-031-17782-8_8

3. ([1]) Simple arguments show that $BA = AB^*$ implies $B^*A = AB$, when B is unitary and A is any (unbounded) operator. If we further assume that A is self-adjoint, then $BA = AB^*$ signifies that BA is self-adjoint and $B^*A = AB$ means that B^*A is self-adjoint. A similar problem is:

Conjecture 8.1.3 Let $B \in B(H)$ be normal and let A be an unbounded self-adjoint operator. Then

$$BA \text{ is self-adjoint} \Longleftrightarrow B^*A \text{ is self-adjoint.}$$

Recall that the previous question has already a positive answer when $A \in B(H)$. However, this is untrue in the case of *two* unbounded operators. See Example 5.4.1.

4. Similarly, we ask:

Conjecture 8.1.4 Let $B \in B(H)$ be normal and let A be an unbounded closed operator. Then

$$BA = AB \Longrightarrow B^*A = AB^*.$$

5. There are more questions to tackle as regards commutativity up to a factor. For instance, is it possible to find two normal operators, A and B, with $B \in B(H)$ such that $BA \subset \lambda AB \neq 0$ with $|\lambda| \neq 1$? One is more tempted to say that this question has a negative answer. In other words, if A and B are normal with $B \in B(H)$, and such that $BA \subset \lambda AB \neq 0$, then we would have $|\lambda| = 1$.

For the "possible proof", if one tries to mimic the proof of the bounded case, then the main issue is the fact that we do not have good tools for handling the situation when A is unbounded. To clarify this point, observe that when $A, B \in B(H)$ are normal and $BA = \lambda AB$, then the Fuglede-Putnam theorem gives

$$BA^* = \bar{\lambda}A^*B \text{ and } \bar{\lambda}AB^* = B^*A,$$

where we have applied it in the first case to the normal operators A and λA, and in the second case to the normal operators B and λB. When $BA \subset \lambda AB$, however, then we only have $BA^* \subset \bar{\lambda}A^*B$. In other words, we cannot a priori have $B^*A \subset \bar{\lambda}AB^*$ as we do not have a version of the Fuglede-Putnam theorem where $B, \lambda B \in B(H)$ are normal (and A is unbounded), as we have already observed. So we conjecture that:

Conjecture 8.1.5 (Cf. [2] and [6]) Let A and B be two normal operators, where $B \in B(H)$. Assume that $BA \subset \lambda AB \neq 0$ with $\lambda \in \mathbb{C}$. Then $B^*A \subset \bar{\lambda}AB^*$.

6. A.E. Nussbaum [10] did not say anything about a possible Putnam's version of Theorem 5.1.1. So, one may ask whether a Fuglede-Putnam version is possible in this case.

7. Regarding Theorem 5.3.5, we conjecture the following:

Conjecture 8.1.6 (Cf. [12], and [4]) Let A and B be two closed hyponormal operators such that $TA^* \subset BT$, where $T \in B(H)$. Then $TA \subset B^*T$.

8. The proof of Theorem 6.4.1 strongly relies on the spectral theorem. I haven't been able to develop an algebraic proof, i.e., "spectral-free". If one can show Embry's theorem without using spectral measures, then one could adapt this possible new proof to generalize Theorem 6.4.1 to some classes of non-normal operators, such as quasinormals, subnormals, and hyponormals.

Conjecture 8.1.7 Let $A, B \in B(H)$ be two commuting hyponormal (resp. subnormal or quasinormal) operators. Let $T \in B(H)$ be such that $0 \notin W(T)$. If $TA = BT$, then $A = B$.

Remark This conjecture first appeared in [7], and it has been unsolved since then.

References

1. I.F.Z. Bensaid, S. Dehimi, B. Fuglede, M.H. Mortad, The Fuglede theorem and some intertwining relations. Adv. Oper. Theor. **6/1**, 8 pp. (2021). Paper No. 9
2. Ch. Chellali, M.H. Mortad, Commutativity up to a factor for bounded and unbounded operators. J. Math. Anal. Appl. **419/1**, 114–122 (2014)
3. S. Dehimi, M.H. Mortad, A. Bachir, Unbounded generalizations of the Fuglede-Putnam theorem. Rend. Istit. Mat. Univ. Trieste **54**(7), 9 (2022). https://doi.org/10.13137/2464-8728/33883
4. R.L. Moore, D.D. Rogers, T.T. Trent, A note on intertwining M-hyponormal operators. Proc. Am. Math. Soc. **83/3**, 514–516 (1981)
5. M.H. Mortad, An application of the Putnam-Fuglede theorem to normal products of self-adjoint operators. Proc. Am. Math. Soc. **131/10**, 3135–3141 (2003)
6. M.H. Mortad, Commutativity up to a factor: more results and the unbounded case. Z. Anal. Anwendungen: Journal for Analysis and its Applications **29/3**, 303–307 (2010)
7. M.H. Mortad, Similarities involving unbounded normal operators. Tsukuba J. Math. **34/1**, 129–136 (2010)
8. M.H. Mortad, *Counterexamples in Operator Theory* (Birkhäuser/Springer, Cham, 2022)
9. M.H. Mortad, Yet another generalization of the Fuglede-Putnam theorem to unbounded operators. arXiv:2003.00339
10. A.E. Nussbaum, A commutativity theorem for unbounded operators in Hilbert space. Trans. Am. Math. Soc. **140**, 485–491 (1969)
11. F.C. Paliogiannis, A generalization of the Fuglede-Putnam theorem to unbounded operators. J. Oper. **2015**, 3 (2015). Art. ID 804353 pp.
12. J. Stochel, An asymmetric Putnam-Fuglede theorem for unbounded operators. Proc. Am. Math. Soc. **129/8**, 2261–2271 (2001)

Appendix A
Bounded and Unbounded Linear Operators

A.1 Bounded Linear Operators

First, some general references for readers' convenience are: [1, 4–8, 12, 13, 15, 17, 20–23, 25], and [27].

Now, we recall quickly and without any proof some of the basic definitions and notions in (bounded) operator theory.

- Let H and K be two Hilbert spaces. Call a linear operator $A : H \to K$ bounded if

$$\|Ax\|_H \leq \alpha \|x\|_K$$

for a certain $\alpha \geq 0$ and all $x \in D(A) \subset H$, where $D(A)$ is a linear subspace of H (which is called the domain of A). Otherwise A is called unbounded. The set of all linear bounded operators from H into K is denoted by $B(H, K)$, where, as is customary, $B(H) := B(H, H)$. If $A, B \in B(H)$ are such that $AB = BA$, then we say that A commutes with B. If I designates the usual identity operator on H, then $A \in B(H)$ is said to be invertible provided there exists a (unique) $B \in B(H)$ such that

$$AB = BA = I,$$

we then set $B = A^{-1}$.
- Every $A \in B(H, K)$ possesses a unique adjoint, noted A^*, where $A^* \in B(K, H)$. If $A \in B(H, K)$, then

$$\ker A = (\operatorname{ran} A^*)^{\perp}, \quad \ker A^* = (\operatorname{ran} A)^{\perp}.$$

Say that $A \in B(H)$ is self-adjoint when $A = A^*$, and call A normal if $AA^* = A^*A$. If $A^*A = I$, then A is an isometry, whereas $AA^* = I$ means that A is a co-isometry. When $AA^* = A^*A = I$, then A is called unitary.

- Any $T \in B(H)$ may be expressed as $A + iB$, where A and B are two bounded self-adjoint operators on H, and vice versa. The decomposition $T = A + iB$ is called the Toeplitz or Cartesian decomposition of $T \in B(H)$.
  The self-adjoint operators

$$\text{Re } T = \frac{T + T^*}{2} \quad \text{and} \quad \text{Im } T = \frac{T - T^*}{2i}$$

are called the real and the imaginary parts of T respectively.
  In the end, recall that $T = A + iB$ is normal iff A commutes with B.

- Let $S \in B(\ell^2)$ be defined by

$$S(x_1, x_2, \cdots) = (0, x_1, x_2, \cdots).$$

Then S is called the (unilateral) shift operator.
It is sometimes more appropriate to define an operator using its effect on the standard (or other) orthonormal basis. For instance, the shift acts on the standard basis $\{e_n : n \in \mathbb{N}\}$ as

$$Se_n = e_{n+1}, \ n \in \mathbb{N}.$$

We may also represent it as

$$S = \begin{pmatrix} 0 & 0 & 0 & 0 & \cdots \\ 1 & 0 & 0 & 0 & \cdots \\ 0 & 1 & 0 & 0 & \cdots \\ 0 & 0 & 1 & 0 & \cdots \\ \vdots & 0 & 0 & 1 & \ddots \\ \vdots & \vdots & \vdots & \ddots & \ddots \end{pmatrix}.$$

The adjoint of the shift operator is given by

$$S^*(x_1, x_2, x_3, \cdots) = (x_2, x_3, \cdots).$$

Its action on the Hilbert basis $\{e_n : n \in \mathbb{N}\}$ is then:

$$S^*e_1 = 0_{\ell^2}, \ S^*e_n = e_{n-1}, \ n \geq 2.$$

On $\ell^2(\mathbb{Z})$, define an operator R by

$$R(x_n)_{n\in\mathbb{Z}} = R(\cdots, x_{-1}, \boldsymbol{x_0}, x_1, \cdots) = (\cdots, x_{-2}, \boldsymbol{x_{-1}}, x_0, \cdots),$$

where the "bold" indicates the zeros positions of the sequence. The operator R is called the bilateral (forward or right) shift.
If R is the bilateral shift on $\ell^2(\mathbb{Z})$, then $\|R\| = 1$ and its adjoint is given by

$$R^*(\cdots, x_{-1}, \boldsymbol{x_0}, x_1, \cdots) = (\cdots, x_{-1}, x_0, \boldsymbol{x_1}, x_2, \cdots).$$

• Let H_1, H_2 be two Hilbert spaces, and let $A_{ij} \in B(H_j, H_i)$, where $i, j = 1, 2$. The matrix

$$A = \begin{pmatrix} A_{11} & A_{12} \\ A_{21} & A_{22} \end{pmatrix}$$

is called an operator matrix. It defines a linear operator from $H_1 \oplus H_2$ into $H_1 \oplus H_2$ by

$$A \begin{pmatrix} x_1 \\ x_2 \end{pmatrix} = \begin{pmatrix} A_{11}x_1 + A_{12}x_2 \\ A_{21}x_1 + A_{22}x_2 \end{pmatrix}$$

Here are some basic properties of operator matrices (borrowed from [15]). Let $A_{ij} \in B(H_j, H_i)$, where $i, j = 1, 2$ and $B_{ij} \in B(H_j, H_i)$, where $i, j = 1, 2$. Set

$$A = \begin{pmatrix} A_{11} & A_{12} \\ A_{21} & A_{22} \end{pmatrix} \text{ and } B = \begin{pmatrix} B_{11} & B_{12} \\ B_{21} & B_{22} \end{pmatrix}.$$

Then

1. $A \in B(H_1 \oplus H_2)$ and

$$\max_{1 \le i,j \le 2} \|A_{ij}\| \le \|A\| \le 4 \max_{1 \le i,j \le 2} \|A_{ij}\|.$$

2. The sum $A + B$ is given by

$$A + B = \begin{pmatrix} A_{11} + B_{11} & A_{12} + B_{12} \\ A_{21} + B_{21} & A_{22} + B_{22} \end{pmatrix}.$$

3. The product AB is given by

$$AB = \begin{pmatrix} A_{11}B_{11} + A_{12}B_{21} & A_{11}B_{12} + A_{12}B_{22} \\ A_{21}B_{11} + A_{22}B_{21} & A_{21}B_{12} + A_{22}B_{22} \end{pmatrix}$$

4. The adjoint A^* of A is given by

$$A^* = \begin{pmatrix} A_{11} & A_{12} \\ A_{21} & A_{22} \end{pmatrix}^* = \begin{pmatrix} A_{11}^* & A_{21}^* \\ A_{12}^* & A_{22}^* \end{pmatrix}.$$

- Let H be a Hilbert space and let (A_n) be a sequence in $B(H)$.

  1. Say that (A_n) converges in norm (or uniformly) to $A \in B(H)$ if

  $$\lim_{n \to \infty} \| A_n - A \| = 0.$$

  The topology associated with this convergence is called the topology of the operator norm (or the uniform topology).

  2. We say that (A_n) strongly converges to $A \in B(H)$ if

  $$\lim_{n \to \infty} \| (A_n - A)x \| = 0$$

  for each $x \in H$. We may then write $s - \lim_{n \to \infty} A_n = A$. The topology associated with this convergence is called the strong operator topology (denoted also by S.O.T.).

  3. We say that (A_n) weakly converges to $A \in B(H)$ if

  $$\lim_{n \to \infty} \langle A_n x, y \rangle = \langle A x, y \rangle$$

  for each $x, y \in H$. We may write $w - \lim_{n \to \infty} A_n = A$. The topology associated with this convergence is called the weak operator topology (denoted also by W.O.T.).

  Recall that

  Convergence in norm $\implies$ Strong convergence $\implies$ Weak convergence.

- Let H be a $\mathbb{C}$-Hilbert space and let $A \in B(H)$. Say that A is positive if $\langle Ax, x \rangle \geq 0$ for all $x \in H$. In symbols, $A \geq 0$. If $-A \geq 0$, then A is said to be negative, and we write $A \leq 0$. If $\langle Ax, x \rangle > 0$ for all $x \in H$ with $x \neq 0$, then A is called strictly positive (noted $A > 0$). Every positive operator on a complex space is self-adjoint. If, however, H is an $\mathbb{R}$-Hilbert, then $A \in B(H)$ is called positive when $A = A^*$ and $\langle Ax, x \rangle \geq 0$ for all $x \in H$. Let $A, B \in B(H)$. Write $A \geq B$ if both A and B are *self-adjoint* and $A - B \geq 0$.

- Let $A \in B(H)$. Say that $B \in B(H)$ is a square root of A provided $B^2 = A$. For instance, an operator may have no square root, as it may have an infinite number of square roots. Nonetheless, a positive operator $A \in B(H)$ has only one positive square root, denoted exclusively by $\sqrt{A}$. If $T \in B(H)$ and A is positive such that $TA = AT$, then $T\sqrt{A} = \sqrt{A}T$.

If $A \in B(H)$, then A^*A is positive. Its unique positive square root, i.e., $\sqrt{A^*A}$, is called the absolute value of A, and it is noted $|A|$. For example, every invertible operator $A \in B(H)$ may be uniquely expressed as $A = U|A|$, where U is unitary.

- Now, we recall two important subsets of $\mathbb{C}$. Let H be a Hilbert space and let $A \in B(H)$.

The set

$$\sigma(A) = \{\lambda \in \mathbb{C} : \lambda I - A \text{ is not invertible}\}$$

is called the spectrum of A. Its complement, that is $\mathbb{C} \setminus \sigma(A)$, is called the resolvent set of A, and it is denoted by $\rho(A)$.

The numerical range of A is defined by

$$W(A) = \{\langle Ax, x \rangle : x \in H, \ \|x\| = 1\}.$$

- Like their finite-dimensional brethren, normal operators on infinite-dimensional Hilbert spaces have a spectral theorem. First, we recall the definition of a spectral measure:

Definition A.1.1 Let X be a non-empty set and let Σ be a sigma algebra of subsets of X. Let H be a Hilbert space. A spectral measure is a function $E : \Sigma \to B(H)$ such that:

1. $E(\varnothing) = 0$ and $E(X) = I$ (I is the identity operator).
2. For every $\Delta \in \Sigma$, $E(\Delta)$ is an orthogonal projection.
3. For any $\Delta_1, \Delta_2 \in \Sigma$: $E(\Delta_1 \cap \Delta_2) = E(\Delta_1)E(\Delta_2)$.
4. If $(\Delta_n)_{n \in \mathbb{N}} \in \Sigma$ are pairwise disjoint, then

$$E\left(\bigcup_{n=1}^{\infty} \Delta_n\right) = \sum_{n=1}^{\infty} E(\Delta_n)$$

w.r.t. the strong operator topology.

Remarks

1. By (3), we have

$$E(\Delta_1)E(\Delta_2) = E(\Delta_2)E(\Delta_1)$$

for any $\Delta_1, \Delta_2 \in \Sigma$. In particular, if $\Delta_1 \cap \Delta_2 = \varnothing$, then

$$E(\Delta_1)E(\Delta_2) = E(\Delta_2)E(\Delta_1) = 0,$$

i.e., $E(\Delta_1)$ and $E(\Delta_2)$ have orthogonal ranges.

2. If $(\Delta_n)_{n\in\mathbb{N}}$ constitutes a partition of X, then (strongly)

$$\sum_{n=1}^{N} E(\Delta_n) \longrightarrow E(X) = I$$

as $N \to \infty$.

Theorem A.1.1 (Spectral Theorem for Normal Operators: Integral Form, [4])
*Let $A \in B(H)$ be normal. Then there exists a unique spectral measure E on the
Borel subsets of $\sigma(A)$ such that:*

$$A = \int_{\sigma(A)} \lambda \, dE,$$

also written as (for all $x, y \in H$)

$$\langle Ax, y \rangle = \int_{\sigma(A)} \lambda \, d\mu_{x,y},$$

where $\mu_{x,y}$ is an ordinary countably additive measure defined by

$$\mu_{x,y}(\Delta) = \langle E(\Delta)x, y \rangle,$$

where $x, y \in H$.
   If Y is a (non-empty) relatively open subset of $\sigma(A)$, then $E(Y) \neq 0$.
   Also,

$$AE(\Delta) = E(\Delta)A$$

*holds for any Borel set Δ and hence $\mathrm{ran}(E(\Delta))$ reduces A. More generally, if $B \in
B(H)$, then*

$$BA = AB \text{ and } BA^* = A^*B \iff BE(\Delta) = E(\Delta)B, \text{ for all } \Delta.$$

Remark The last statement of the preceding theorem is the original statement of the
spectral theorem, and we already know that $BA = AB$ if and only if $BA^* = A^*B$
thanks to the Fuglede theorem.

Remark If f is bounded and Borel on $\sigma(A)$, then define

$$f(A) = \int_{\sigma(A)} f(\lambda) \, dE$$

where E is the spectral measure associated with A.

Corollary A.1.2 *If A, $B \in B(H)$ are both normal with E and F as the corresponding spectral measures, then*

$$AB = BA \iff E(\Delta)F(\Delta') = F(\Delta')E(\Delta)$$

for all Borel sets Δ and Δ'.

- As a consequence of the spectral theorem, a normal operator with a real spectrum is self-adjoint. See [16] for a new proof.
- Let $T \in B(H)$ be fixed, and define the linear operator

$$X \mapsto \Delta_T(X) = TX - XT$$

on $B(H)$. Then $\Delta_T \in B[B(H)]$ is called a derivation.

Let us say a word why such an operator is called a derivation. To this end, let $X, Y \in B(H)$. Then

$$\Delta_T(XY) = T(XY) - (XY)T = \Delta_T(X)Y + X\Delta_T(Y),$$

and so we get something that resembles the usual product rule for derivatives.

If $T, S \in B(H)$ are fixed, then the bounded linear operator $X \mapsto \Delta_{T,S}(X) = TX - XS$, defined on $B(H)$, is called a generalized derivation.

- We finish this section by recalling some classes of non-normal operators. Other classes will be defined in due time. Let $A \in B(H)$. Say that A is hyponormal if

$$\forall x \in H : \|A^*x\| \leq \|Ax\|.$$

Obviously, every normal operator is hyponormal. The shift is the prominent example of a hyponormal operator that is not normal. Clearly

$$A \text{ is hyponormal} \iff AA^* \leq A^*A.$$

We say that the operator A is co-hyponormal if A^* is hyponormal, that is

$$\forall x \in H : \|Ax\| \leq \|A^*x\|.$$

Equivalently, this amounts to say that $AA^* \geq A^*A$. Obviously, $A \in B(H)$ is normal *if and only if* A is both hyponormal *and* co-hyponormal.

We say that A is subnormal if it possesses a normal extension $N \in B(K)$ (with $N(H) \subset H$) where K is a Hilbert space larger than H, i.e., $H \subset K$. In other words, $A \in B(H)$ is subnormal if $K = H \oplus H^\perp$ and $N \in B(K)$, defined by

$$N = \begin{pmatrix} A & B \\ 0 & C \end{pmatrix},$$

is normal for some $B \in B(H^\perp, H)$ and $C \in B(H^\perp)$.

Say that A is quasinormal if

$$A(A^*A) = (A^*A)A.$$

Finally, we say that A is paranormal if

$$\|Ax\|^2 \le \|A^2x\|$$

for any *unit* vector $x \in H$. Equivalently, $\|Ax\|^2 \le \|A^2x\| \|x\|$ for any $x \in H$.

Examples A.1.3

1. Each normal operator is quasinormal.
2. The shift operator is quasinormal. In fact, each isometry is quasinormal.

We have the following inclusions among those classes (see, e.g., [9] for a proof):

Quasinormal $\subset$ Subnormal $\subset$ Hyponormal $\subset$ Paranormal $\subset$ Normaloid.

When dim $H < \infty$, subnormality, paranormality, quasinormality and hyponormality all coincide with normality.

A.2   Unbounded Linear Operators

This section is devoted to basic definitions and notions about unbounded operators. Some references are [2–4, 6, 17, 19, 20, 22, 24], and [25].

Let H and K be two Hilbert spaces.

- Say that a linear operator B, with a domain $D(B)$, is an extension of another linear operator A, with a domain $D(A)$, and we write $A \subset B$, if

$$D(A) \subset D(B) \text{ and } \forall x \in D(A): \ Ax = Bx.$$

- Let $A : D(A) \subset H \to K$ be a densely defined linear operator, i.e., $\overline{D(A)} = H$. Then $D(A^*)$ is constituted of all $y \in K$ for which there exists a $z \in H$ such that

$$\langle Ax, y \rangle = \langle x, z \rangle, \; \forall x \in D(A).$$

The (linear) operator A^* then obtained is called the adjoint of A.
- A linear operator A is called closed provided its graph is closed in $H \oplus K$. It is called closable if it has a closed extension. The smallest closed extension of it is called its closure and it is denoted by $\overline{A}$. A standard result states that A is closable iff A^* has a dense domain, and in which case $\overline{A} = A^{**}$. Also, if A is closable, then $(\overline{A})^* = A^*$.
  If A is closable and B is a linear operator, and $B \subset A$, then B is closable and $\overline{B} \subset \overline{A}$.
- Let A be a densely defined linear operator. Say that A is symmetric provided that $A \subset A^*$ (notice that symmetric operators may be defined on a non-dense subspace $D(A)$ by requiring $\langle Ax, y \rangle = \langle x, Ay \rangle$ for all $x, y \in D(A)$). It is called self-adjoint if $A = A^*$, i.e., if A is symmetric and $D(A) = D(A^*)$. If $\overline{A}$ is self-adjoint, then A is called essentially self-adjoint.
- If A and B are two operators with domains $D(A)$ and $D(B)$ respectively, then the product AB is defined by

$$(AB)x := A(Bx)$$

for x in the domain

$$D(AB) = \{x \in D(B) : \; Bx \in D(A)\} = B^{-1}[D(A)].$$

Similarly, the sum $A + B$ is defined as

$$(A + B)x := Ax + Bx$$

for x in the domain

$$D(A + B) = D(A) \cap D(B).$$

- Say that a densely defined linear operator A is normal when A is closed and $AA^* = A^*A$, or equivalently, A is normal if and only if $\|Ax\| = \|A^*x\|$ for all $x \in D(A) = D(A^*)$. Observe that if A is normal, then so is A^*. But, if A^* is normal, then A is normal if it is closed.
- Self-adjoint operators are maximally symmetric, that is, if A is self-adjoint and B is symmetric, then

$$A \subset B \Longrightarrow A = B.$$

Normal operators too are maximally normal.

- We say that $B \in B(H)$ commutes with a linear operator A with domain $D(A) \subset H$ when $BA \subset AB$, that is when $BAx = ABx$ for all $x \in D(A) \subset D(AB)$.
- Let A be an injective operator from $D(A)$ into H. Then $A^{-1} : \text{ran}(A) \to H$ is called the inverse of A, with $D(A^{-1}) = \text{ran}(A)$.

  If the inverse of an unbounded operator is bounded and everywhere defined (e.g., if $A : D(A) \to H$ is closed and bijective), then A is said to be boundedly invertible. In other words, such is the case if there is a $B \in B(H)$ such that

$$AB = I \text{ and } BA \subset I.$$

  If A is boundedly invertible, then it is closed. Recall also that $T + S$ is closed (resp. closable) if $S \in B(H)$ and T is closed (resp. closable), and that ST is closed (resp. closable) if, e.g., S is closed (resp. closable) and $T \in B(H)$. If S is boundedly invertible and T is closed, then ST is closed as well.

- The spectrum of unbounded operators is defined as follows: Let A be an operator. The resolvent set of A, denoted by $\rho(A)$, is defined by

$$\rho(A) = \{\lambda \in \mathbb{C} : \lambda I - A \text{ is bijective and } (\lambda I - A)^{-1} \in B(H)\}.$$

  The complement of $\rho(A)$, denoted by $\sigma(A)$, i.e.,

$$\sigma(A) = \mathbb{C} \setminus \rho(A)$$

  is called the spectrum of A.

  Clearly, $\lambda \in \rho(A)$ iff there is a $B \in B(H)$ such that

$$(\lambda I - A)B = I \text{ and } B(\lambda I - A) \subset I.$$

  If $\sigma(A) \neq \mathbb{C}$, then A is necessarily closed.

- Let A and B be two densely defined linear operators defined on a Hilbert space H.

  1. If $A + B$ is densely defined, then $A^* + B^* \subset (A + B)^*$.
  2. If $B \in B(H)$, then $A^* + B^* = (A + B)^*$.
  3. If BA is densely defined, then $A^*B^* \subset (BA)^*$.
  4. If $B \in B(H)$, then $A^*B^* = (BA)^*$.
  5. If BA is densely defined, and A is invertible and such that $A^{-1} \in B(H)$, then also $A^*B^* = (BA)^*$.

- If A is a densely defined closed operator, then both AA^* and A^*A are self-adjoint.
- As in the case of bounded self-adjoint or normal operators, we have different forms of the spectral theorem for unbounded self-adjoint or normal operators.

Theorem A.2.1 (Spectral Theorem for Unbounded Normal Operators: Integral Form) *Let A be a normal with domain $D(A) \subset H$. Then there exists a unique spectral measure E on the Borel subsets of $\mathbb{C}$ such that:*

$$A = \int_{\sigma(A)} \lambda dE,$$

If f is measurable and finite almost everywhere, then define

$$f(A) = \int_{\sigma(A)} f(\lambda)dE.$$

Readers may wish to consult Chaps. 4 and 5 in [24] for an exhaustive treatment of the spectral theorem as well as the functional calculus of unbounded self-adjoint or normal operators.

- An unbounded normal operator with a real spectrum is self-adjoint.
- Two (unbounded) self-adjoint (or normal) operators A and B are said to strongly commute if their corresponding spectral measures commute.
- Let A, B be two self-adjoint (or normal) operators.

  (i) If $A, B \in B(H)$, then $AB = BA$ iff A and B strongly commute.
  (ii) If $B \in B(H)$, then $BA \subset AB$ iff A and B strongly commute.

- A symmetric operator A is called positive if

$$\langle Ax, x \rangle \geq 0, \forall x \in D(A).$$

Every unbounded self-adjoint positive A has a unique self-adjoint positive square root, denoted by $\sqrt{A}$ or $A^{1/2}$. The proof is based upon the spectral theorem. The absolute value of a densely defined closed A is given by $|A| = \sqrt{A^*A}$. It is also known that $D(A) = D(|A|)$ when A is closed and densely defined.

- Let us now recall some rudimentary facts about matrices of non-necessarily bounded operators. Let H and K be two Hilbert spaces and let $A : H \oplus K \to H \oplus K$ (we may also use $H \times K$ instead of $H \oplus K$) be defined by

$$A = \begin{pmatrix} A_{11} & A_{12} \\ A_{21} & A_{22} \end{pmatrix} \tag{A.1}$$

where $A_{11} \in L(H)$, $A_{12} \in L(K, H)$, $A_{21} \in L(H, K)$ and $A_{22} \in L(K)$ are not necessarily bounded operators. If A_{ij} has a domain $D(A_{ij})$ with $i, j = 1, 2$, then

$$D(A) = (D(A_{11}) \cap D(A_{21})) \times (D(A_{12}) \cap D(A_{22}))$$

is the natural domain of A. So if $(x_1, x_2) \in D(A)$, then

$$A \begin{pmatrix} x_1 \\ x_2 \end{pmatrix} = \begin{pmatrix} A_{11}x_1 + A_{12}x_2 \\ A_{21}x_1 + A_{22}x_2 \end{pmatrix}.$$

As is customary, we allow the abuse of notation $A(x_1, x_2)$. Readers should also be careful when dealing with products of matrices of (unbounded) operators as they may encounter some issues with their domains.

Also, recall that the adjoint of $\begin{pmatrix} A_{11} & A_{12} \\ A_{21} & A_{22} \end{pmatrix}$ is not always equal to $\begin{pmatrix} A_{11}^* & A_{21}^* \\ A_{12}^* & A_{22}^* \end{pmatrix}$

(even when all domains are dense including the main domain $D(A)$) as known counterexamples show. Nonetheless, e.g.,

$$\begin{pmatrix} A & 0 \\ 0 & B \end{pmatrix}^* = \begin{pmatrix} A^* & 0 \\ 0 & B^* \end{pmatrix}$$

and

$$\begin{pmatrix} 0 & C \\ D & 0 \end{pmatrix}^* = \begin{pmatrix} 0 & D^* \\ C^* & 0 \end{pmatrix}$$

if A, B, C and D are all densely defined. See, e.g., [26] for proofs of other results, and for more about unbounded operator matrices.

- In the end, we recall some definitions of unbounded non-normal operators. A densely defined operator A with domain $D(A)$ is called hyponormal if

$$D(A) \subset D(A^*) \text{ and } \|A^*x\| \leq \|Ax\|, \ \forall x \in D(A).$$

A densely defined linear operator A with domain $D(A) \subset H$, is said to be subnormal when there are a Hilbert space K with $H \subset K$, and a normal operator N with $D(N) \subset K$ such that

$$D(A) \subset D(N) \text{ and } Ax = Nx \text{ for all } x \in D(A).$$

A quasinormal operator A is a closed densely defined one such that $AA^*A = A^*A^2$ (or $AA^*A \subset A^*A^2$ as in say [10]).

A densely defined A is said to be formally normal provided

$$\|Ax\| = \|A^*x\|, \ \forall x \in D(A) \subset D(A^*).$$

Remark The notion of paranormality also exists for non-necessarily bounded operators (see, e.g., [18]).

As in the bounded case, we have (see, e.g., [10, 11, 14] and J. Stochel and F.H. Szafraniec (Unbounded operators and subnormality, unpublished yet)):

$$\text{Normal} \subsetneq \text{Quasinormal} \subsetneq \text{Subnormal} \subsetneq \text{Hyponormal}.$$

References

1. S.K. Berberian, *Introduction to Hilbert Space*. Reprinting of the 1961 original. With an addendum to the original (Chelsea Publishing Co., New York, 1976)
2. Y.M. Berezansky, Z.G. Sheftel, G.F. Us, *Functional Analysis*, vol. II. Translated from the 1990 Russian original by Peter V. Malyshev. Operator Theory: Advances and Applications, vol. 86 (Birkhäuser, Basel, 1996)
3. M.Sh. Birman, M.Z. Solomjak, *Spectral Theory of Selfadjoint Operators in Hilbert Space*. Translated from the 1980 Russian original by S. Khrushchëv and V. Peller. Mathematics and its Applications (Soviet Series) (D. Reidel Publishing Co., Dordrecht, 1987)
4. J.B. Conway, *A Course in Functional Analysis*, 2nd edn. (Springer, New York, 1990)
5. J.B. Conway, *The Theory of Subnormal Operators*, vol. 36. Mathematical Surveys and Monographs (*American Mathematical Society*, Providence, RI, 1991)
6. J. Dieudonné, *Treatise on Analysis*, vol. II. Enlarged and corrected printing. Translated by I.G. Macdonald. With a loose erratum. Pure and Applied Mathematics, 10-II (Academic Press, [Harcourt Brace Jovanovich, Publishers], New York, London, 1976)
7. N. Dunford, J.T. Schwartz, *Linear Operators. Part I. General Theory*. With the assistance of William G. Bade and Robert G. Bartle. Reprint of the 1958 original. Wiley Classics Library. A Wiley-Interscience Publication (Wiley, New York, 1988)
8. N. Dunford, J.T. Schwartz, *Linear operators. Part II. Spectral theory. Selfadjoint operators in Hilbert space*. With the assistance of William G. Bade and Robert G. Bartle. Reprint of the 1963 original. Wiley Classics Library. A Wiley-Interscience Publication (Wiley, New York, 1988)
9. T. Furuta, *Invitation to Linear Operators: From Matrices to Bounded Linear Operators on a Hilbert Space* (Taylor & Francis, Ltd., London, 2001)
10. Z.J. Jabłoński, I.B. Jung, J. Stochel, Unbounded quasinormal operators revisited. Integr. Equ. Oper. Theor. **79/1**, 135–149 (2014)
11. J. Janas, On unbounded hyponormal operators. Ark. Mat. **27/2**, 273–281 (1989)
12. R.V. Kadison, J.R. Ringrose, *Fundamentals of the Theory of Operator Algebras*, vol. I. Elementary theory. Reprint of the 1983 original, G.S.M., vol. **15** (American Mathematical Society, Providence, RI, 1997)
13. C.S. Kubrusly, *The Elements of Operator Theory*, 2nd edn. (Birkhäuser/Springer, New York, 2011)
14. W. Majdak, A lifting theorem for unbounded quasinormal operators. J. Math. Anal. Appl. **332/2**, 934–946 (2007)
15. M.H. Mortad, *An Operator Theory Problem Book* (World Scientific Publishing Co., Singapore, 2018)
16. M.H. Mortad, On the invertibility of the sum of operators. Anal. Math. **46/1**, 133–145 (2020)
17. M.H. Mortad, *Counterexamples in Operator Theory* (Birkhäuser/Springer, Cham, 2022)
18. M.H. Mortad, Counterexamples related to unbounded paranormal operators. *Examples and Counterexamples* (to appear). https://doi.org/10.1016/j.exco.2021.100017
19. G.K. Pedersen, Analysis now. *Graduate Texts in Mathematics*, vol. 118 (Springer, New York, 1989)
20. M. Reed, B. Simon, *Methods of Modern Mathematical Physics*, vol. 1. Functional analysis (Academic Press, New York, 1980)

21. F. Riesz, B. Sz.-Nagy, *Functional Analysis*. Translated from the second French edition by Leo F. Boron. Reprint of the 1955 original. Dover Books on Advanced Mathematics (Dover Publications, Inc., New York, 1990)
22. W. Rudin, *Functional Analysis*, 2nd edn. (McGraw-Hill, New York, 1991)
23. R. Schatten, Norm ideals of completely continuous operators. Second printing. Ergebnisse der Mathematik und ihrer Grenzgebiete, Band **27** (Springer, Berlin, New York, 1970)
24. K. Schmüdgen, *Unbounded Self-adjoint Operators on Hilbert Space*, vol. 265. GTM (Springer, New York, 2012)
25. B. Simon, *Operator Theory. A Comprehensive Course in Analysis*, Part 4 (American Mathematical Society, Providence, RI, 2015)
26. Ch. Tretter, *Spectral Theory of Block Operator Matrices and Applications* (Imperial College Press, London, 2008)
27. J. Weidmann, *Linear Operators in Hilbert Spaces* (translated from the German by J. Szücs). GTM, vol. 68 ((Springer, New York, 1980)

Index

© The Author(s), under exclusive license to Springer Nature Switzerland AG 2022
M. H. Mortad, *The Fuglede-Putnam Theory*, Lecture Notes in Mathematics 2322,
https://doi.org/10.1007/978-3-031-17782-8

LECTURE NOTES IN MATHEMATICS

Editors in Chief: J.-M. Morel, B. Teissier;

Editorial Policy

1. Lecture Notes aim to report new developments in all areas of mathematics and their applications – quickly, informally and at a high level. Mathematical texts analysing new developments in modelling and numerical simulation are welcome.

   Manuscripts should be reasonably self-contained and rounded off. Thus they may, and often will, present not only results of the author but also related work by other people. They may be based on specialised lecture courses. Furthermore, the manuscripts should provide sufficient motivation, examples and applications. This clearly distinguishes Lecture Notes from journal articles or technical reports which normally are very concise. Articles intended for a journal but too long to be accepted by most journals, usually do not have this "lecture notes" character. For similar reasons it is unusual for doctoral theses to be accepted for the Lecture Notes series, though habilitation theses may be appropriate.

2. Besides monographs, multi-author manuscripts resulting from SUMMER SCHOOLS or similar INTENSIVE COURSES are welcome, provided their objective was held to present an active mathematical topic to an audience at the beginning or intermediate graduate level (a list of participants should be provided).

   The resulting manuscript should not be just a collection of course notes, but should require advance planning and coordination among the main lecturers. The subject matter should dictate the structure of the book. This structure should be motivated and explained in a scientific introduction, and the notation, references, index and formulation of results should be, if possible, unified by the editors. Each contribution should have an abstract and an introduction referring to the other contributions. In other words, more preparatory work must go into a multi-authored volume than simply assembling a disparate collection of papers, communicated at the event.

3. Manuscripts should be submitted either online at www.editorialmanager.com/lnm to Springer's mathematics editorial in Heidelberg, or electronically to one of the series editors. Authors should be aware that incomplete or insufficiently close-to-final manuscripts almost always result in longer refereeing times and nevertheless unclear referees' recommendations, making further refereeing of a final draft necessary. The strict minimum amount of material that will be considered should include a detailed outline describing the planned contents of each chapter, a bibliography and several sample chapters. Parallel submission of a manuscript to another publisher while under consideration for LNM is not acceptable and can lead to rejection.

4. In general, **monographs** will be sent out to at least 2 external referees for evaluation.

   A final decision to publish can be made only on the basis of the complete manuscript, however a refereeing process leading to a preliminary decision can be based on a pre-final or incomplete manuscript.

   Volume Editors of **multi-author works** are expected to arrange for the refereeing, to the usual scientific standards, of the individual contributions. If the resulting reports can be forwarded to the LNM Editorial Board, this is very helpful. If no reports are forwarded or if other questions remain unclear in respect of homogeneity etc, the series editors may wish to consult external referees for an overall evaluation of the volume.

5. Manuscripts should in general be submitted in English. Final manuscripts should contain at least 100 pages of mathematical text and should always include

   - a table of contents;
   - an informative introduction, with adequate motivation and perhaps some historical remarks: it should be accessible to a reader not intimately familiar with the topic treated;
   - a subject index: as a rule this is genuinely helpful for the reader.
   - For evaluation purposes, manuscripts should be submitted as pdf files.

6. Careful preparation of the manuscripts will help keep production time short besides ensuring satisfactory appearance of the finished book in print and online. After acceptance of the manuscript authors will be asked to prepare the final LaTeX source files (see LaTeX templates online: https://www.springer.com/gb/authors-editors/book-authors-editors/manuscriptpreparation/5636) plus the corresponding pdf- or zipped ps-file. The LaTeX source files are essential for producing the full-text online version of the book, see http://link.springer.com/bookseries/304 for the existing online volumes of LNM). The technical production of a Lecture Notes volume takes approximately 12 weeks. Additional instructions, if necessary, are available on request from lnm@springer.com.

7. Authors receive a total of 30 free copies of their volume and free access to their book on SpringerLink, but no royalties. They are entitled to a discount of 33.3 % on the price of Springer books purchased for their personal use, if ordering directly from Springer.

8. Commitment to publish is made by a *Publishing Agreement*; contributing authors of multiauthor books are requested to sign a *Consent to Publish form*. Springer-Verlag registers the copyright for each volume. Authors are free to reuse material contained in their LNM volumes in later publications: a brief written (or e-mail) request for formal permission is sufficient.

Addresses:
Professor Jean-Michel Morel, CMLA, École Normale Supérieure de Cachan, France
E-mail: moreljeanmichel@gmail.com

Professor Bernard Teissier, Equipe Géométrie et Dynamique,
Institut de Mathématiques de Jussieu – Paris Rive Gauche, Paris, France
E-mail: bernard.teissier@imj-prg.fr

Springer: Ute McCrory, Mathematics, Heidelberg, Germany,
E-mail: lnm@springer.com

Printed in the United States
by Baker & Taylor Publisher Services